Ecological Diversity
and Its Measurement

Ecological Diversity and Its Measurement

Anne E. Magurran

PRINCETON UNIVERSITY PRESS
Princeton, New Jersey

First published 1988 by Princeton University Press,
41 William Street, Princeton, New Jersey 08540

Library of Congress Cataloging in Publication Data

Magurran, Anne E.
Ecological diversity and its measurement / Anne E. Magurran.
p. cm.
Bibliography: p.
Includes index.
ISBN 0-691-08485-8 ISBN 0-691-08491-2 (pbk.)
1. Biological diversity—Measurement. I. Title.
QH75.M32 1988
574.5′248—dc19 88—10927
CIP

Printed in the United States of America
by Princeton University Press, Princeton, N.J., 08540

9 8 7 6 5 4 3 2

To my parents
Roy and Mary Magurran

Contents

Preface

Although diversity is one of the central themes of ecology there is considerable disagreement about how it should be measured. I first encountered this problem 10 years ago when I started my research career and spent a long time pouring over the literature in order to find the most useful techniques. The intervening decade has seen a further increase in the number of papers devoted to the topic of ecological diversity but has led to no consensus on how it should be measured. My aim in writing this book is therefore to provide a practical guide to ecological diversity and its measurement. In a quantitative subject such as the measurement of diversity it is inevitable that some mathematics are involved, but at all times these are kept as simple as possible, and the emphasis is constantly on ecological reality and practical application. I hope that others entering the fascinating field of ecological diversity will find it helpful.

This book grew out of my work in The School of Biological and Environmental Studies at the New University of Ulster, Coleraine, Northern Ireland. I am indebted to all the ecologists there for providing a stimulating atmosphere. Foremost among these were Amyan Macfadyen and Palmer Newbould. A number of the figures and tables in the book are based on data collected in Northern Irish woodlands. It is a pleasure to thank the Northern Ireland Forest Service and Conservation Branch for access to their forests and reserves. I am particularly grateful in this respect to Joe Furphy and John Greer.

Writing a book on diversity and its measurement is rather like setting out across an ecological minefield and I am therefore indebted to the many people who provided advice and ideas. These include Keith Day, Bob May, Ralph Oxley, Stuart Pimm, Tony Pitcher, Brian Rushton and two anonymous referees. The reviewers made helpful and extensive comments on the manuscript. I have incorporated many of their suggestions and feel that the book has been greatly improved by them. The reviewers did not always agree with each other and I am sure that not all readers will approve of my approach! The emphasis and opinions of the book, and any errors that remain, are of course my own responsibility.

Unpublished manuscripts were kindly provided by John Gray, Paul Harvey, Howard Platt, Deborah Rabinowitz and Richard Shattock.

I am grateful to Jane ap Thomas for drawing the vignettes and the pictures of moths, mosses and other organisms.

Finally, I would like to record my thanks to Judith May (Princeton University Press) and Tim Hardwick (Croom Helm).

Anne Magurran,
Bangor 1987

1
Why diversity?

There are three reasons why ecologists are interested in ecological diversity and its measurement. First, despite changing fashions and preoccupations, diversity has remained a central theme in ecology. The well documented patterns of spatial and temporal variation in diversity which intrigued the early investigators of the natural world (for example Clements, 1916; Thoreau, 1860) continue to stimulate the minds of ecologists today (Currie and Paquin, 1987; May, 1986). Second, measures of diversity are frequently seen as indicators of the wellbeing of ecological systems. Thirdly, considerable debate surrounds the measurement of diversity. Diversity may appear to be a straightforward concept which can be quickly and painlessly measured. This is because most people have a ready intuitive grasp of what is meant by diversity and have little difficulty in accepting, say, that tropical rain forests are more diverse than temperate woodlands or that there is a high diversity of organisms in coral reefs. Yet diversity is rather like an optical illusion. The more it is looked at, the less clearly defined it appears to be and viewing it from different angles can lead to different perceptions of what is involved. The problem has been exacerbated by the fact that ecologists have devised a huge range of indices and models for measuring diversity. Despite, or perhaps as a result of these, diversity has a knack of eluding definition and in one instance Hurlbert (1971) even went so far as to decry it as a 'non-concept'.

There is however a simple explanation why diversity is so hard to define. That is because diversity consists of not one but two components. These are first the variety and secondly the relative abundance of species. Table 1.1 lists typical species abundance data and illustrates the way in which the number of species (often referred to as species richness) and their relative abundances can vary. The precise way in which these two factors are incorporated into diversity measures will be elaborated in Chapter 2. It is sufficient for now to note that diversity can be measured by recording the number of species, by describing their relative abundances or by using a measure which combines the two components.

It is important that ecologists should understand how to measure diversity and what they mean by it. Diversity lies at the root of some of the most

Table 1.1 The species diversity of stream insects on *Fontanalis* spp. moss substrate compared to diversity on artificial substrates. These data (taken from Glime and Clemons, 1972) were collected to determine the role of bryophytes as a habitat for stream insects. They contrast the abundance (number of individuals) and variety of species found on real and artificial (plastic and string) mosses.

Substrate	*Moss*	*String*	*Plastic*
Chironimidae	1095	285	190
Simulidae			
Prosimulium hirtipes	111	5	40
Cnephia mutata	82	23	40
Prosimulium rhizophorum	2		2
Nemouridae			
Nemoura sp 4	84	67	10
Nemoura nr. *venosa*	4	7	1
Hydroptilidae			
Agraylea sp. 1	34	2	2
Rhyacophilidae			
Rhyacophila nr. *invaria*	23	10	4
Limnephilidae			
Ironoquia punctatissima	18	5	1
Capniidae			
Allocapnia spp.	17	10	1
Ephemerellidae			
Ephemerella deficiens	12	2	
Ephemerella funeralis	2		
Perlodidae			
Isoperla bilineata	12	1	
Carabidae sp.	11	1	
Veliidae			
Microvelia sp. 3	7	1	1
Lepidostomidae			
Lepidostoma sp. 1	5	1	
Leptophlebiidae			
Leptophlebia sp. 1	5	2	2
Odontoceridae			
Psilotreta frontalis	4		
Hydropsychidae			
Parapsyche apicalis	4	1	1
Helidae			
Bezzia sp. 1	2		
Hydroptilidae			
Paleagapetus celsus	1		
Rhyphidae? sp. 1	1	1	
Baetidae			
Baetis sp. 5	1	1	

Table 1.1—*continued*

Substrate	*Moss*	*String*	*Plastic*
Philopotamidae			
Wormaldia sp. 1	1		
Elmidae			
Promoresia elegans	1		
Isotomidae			
Isotomuros sp. 1		2	
Psychomyiidae			
Polycentropus sp. 1		2	
Hydrophilidae sp. 2		1	
Tipulidae			
Limonia sp. 2		1	
Staphylinadae sp. 1		1	

fundamental and exciting questions in theoretical and applied ecology. For instance, a great deal of effort has been devoted to explaining why there are systematic and predictable latitudinal patterns of diversity (Pianka, 1983; Krebs, 1985; Begon *et al.*, 1986) and why diversity is so closely associated with area (MacArthur and Wilson, 1967; Williamson, 1981). The diversity–stability debate (Elton, 1958; May, 1973, 1981, 1984; Pimm, 1982, 1984) is another example of the ways in which the strands of theoretical and applied ecology intertwine providing rich opportunities for ecologists to further their understanding of the natural world. This book does not set out to provide a discussion of ecological diversity *per se*. Rather, its purpose is to convince ecologists that there are many instances in which it is useful and informative to measure diversity, to provide a guide to the multitude of methods that exist for doing so, and to give advice on the selection and interpretation of diversity measures.

Investigations of ecological diversity are often restricted to species richness, that is a straightforward count of the number of species present. There is however much to interest the ecologist in the relative abundances of species. No community consists of species of equal abundance. Instead, as Table 1.1 shows, and we shall see in more detail in Chapter 2, it is normally the case that the majority of species are rare while a number are moderately common with the remaining few species being very abundant indeed. A variety of species abundance distributions have been proposed to describe the observed patterns (Chapter 2). For instance, in large species-rich communities the distribution of species abundances is usually log normal while in species-poor communities under a harsh environmental regime a geometric series often pertains. Nevertheless, as with the latitudinal gradient of diversity, it is much easier to describe a pattern than to explain it. A number of resource-apportioning theories have been advanced but there is still no concensus about the rules that determine community structure. In fact, there is even a view that the ubiquity of the log normal is an artifact of the mathematics of large data sets. Chapter 2 reviews this and other more biological explanations. The lack of agreement does not however mean that knowledge of species-abundance relationships has no practical value. Environmental monitoring (Chapter 6) makes use of the fact that polluted or stressed communities are characterized by a change in their species abundances which often switch from being log normally distributed to following a geometric series.

Although many branches of ecology are involved with the concept of diversity, in most cases the procedures for measuring diversity are glossed over. This book therefore provides practical advice on the measurement of ecological diversity. It begins with a review of the many diversity indices, models and distributions. Worked examples of the most widely used methods are included because, as Pielou (1984) observes, 'unless one understands a technique, one cannot intelligently judge the results'.

Sampling is another important consideration in studies of ecological diversity. Chapter 3 provides guidance on how to choose the correct sample size, define the study area and select the appropriate technique for measuring abundance.

With so many methods to choose from it can sometimes be difficult to decide which is the most suitable way of measuring diversity. Chapter 4 assesses the performance of a large range of diversity indices according to a set of criteria which include discrimination ability and sensitivity to sample size. It concludes with a guide to the analysis and interpretation of diversity data.

So far this introduction has treated species diversity as being synonymous with ecological diversity. But species diversity is not the only variety of ecological diversity. For instance measures of niche width describe the diversity of resources that an organism (or species) utilizes. Similarly, habitat diversity is an index which measures the structural complexity of the environment or the number of communities present. Methods of measuring niche width and habitat diversity are closely allied to techniques for measuring species diversity. By contrast a rather different approach is adopted when beta (β) diversity is being described. β diversity is defined as the degree of change in (species) diversity along a transect or between habitats. These other varieties of ecological diversity are reviewed in Chapter 5.

The final, and sometimes the most difficult, task for a proponent of diversity measures is to convince fellow ecologists why they should use them. Species richness may only be one component of diversity but it is relatively simple to measure and has been used successfully in many studies. Yet species diversity measures are often more informative than species counts alone. The book therefore concludes with a discussion of the empirical value of diversity measures. It does so in the context of two areas of application. In one of these, environmental monitoring, diversity measures are widely used and have been extensively tested. In the other, conservation management, great score is set on maximizing diversity, which in almost all cases is defined as species richness. Environmental monitoring proves that diversity measures can be empirically useful. Do such measures have an unrealized potential in conservation management? Chapter 6 addresses this question.

2
Diversity indices and species abundance models

A quick dip into the literature on diversity reveals a bewildering range of indices. Each of these indices seeks to characterize the diversity of a sample or community by a single number. To add yet more confusion an index may be known by more than one name and written in a variety of notations using a range of log bases. This diversity of diversity indices has arisen because, for a number of years, it was standard practice for an author to review existing indices, denounce them as useless, and promptly invent a new index. Southwood (1978) notes an interesting parallel in the proliferation of new designs of light traps and new permutations of diversity measures.

On first inspection diversity appears to be a very simple and unambiguous concept. Where then is there scope for so many competing indices? The answer lies in the fact that diversity measures takes into account two factors: species richness, that is number of species, and evenness (sometimes known as equitability), that is how equally abundant the species are. High evenness, which occurs when species are equal or virtually equal in abundance, is conventionally equated with high diversity. These dual concepts of species richness and abundance are illustrated in Figure 2.1. In a comparison between A and B site A would be considered to be more diverse since it has three species of moths (therefore greater richness) while site B has only one. By contrast there is no difference in the species richness of C and D. Site C has four species of moths each with three individuals. Site D also has four species of moths and again a total of 12 individuals. However in the case of site D one species is particularly abundant with nine individuals, the remainder rare with only one individual each. So although C and D have equal numbers of species and individuals the greater evenness of C makes it the more diverse. These examples are of course very simplistic and as we shall soon see situations where species are equally abundant are not a characteristic of the real world. Nevertheless they serve as an introduction to the two concepts which underpin the measurement of diversity. Many of the differences between indices lie in the relative weighting that they give to evenness and species richness.

Species diversity measures can be divided into three main categories. First are the *species richness indices*. These indices are essentially a measure of the

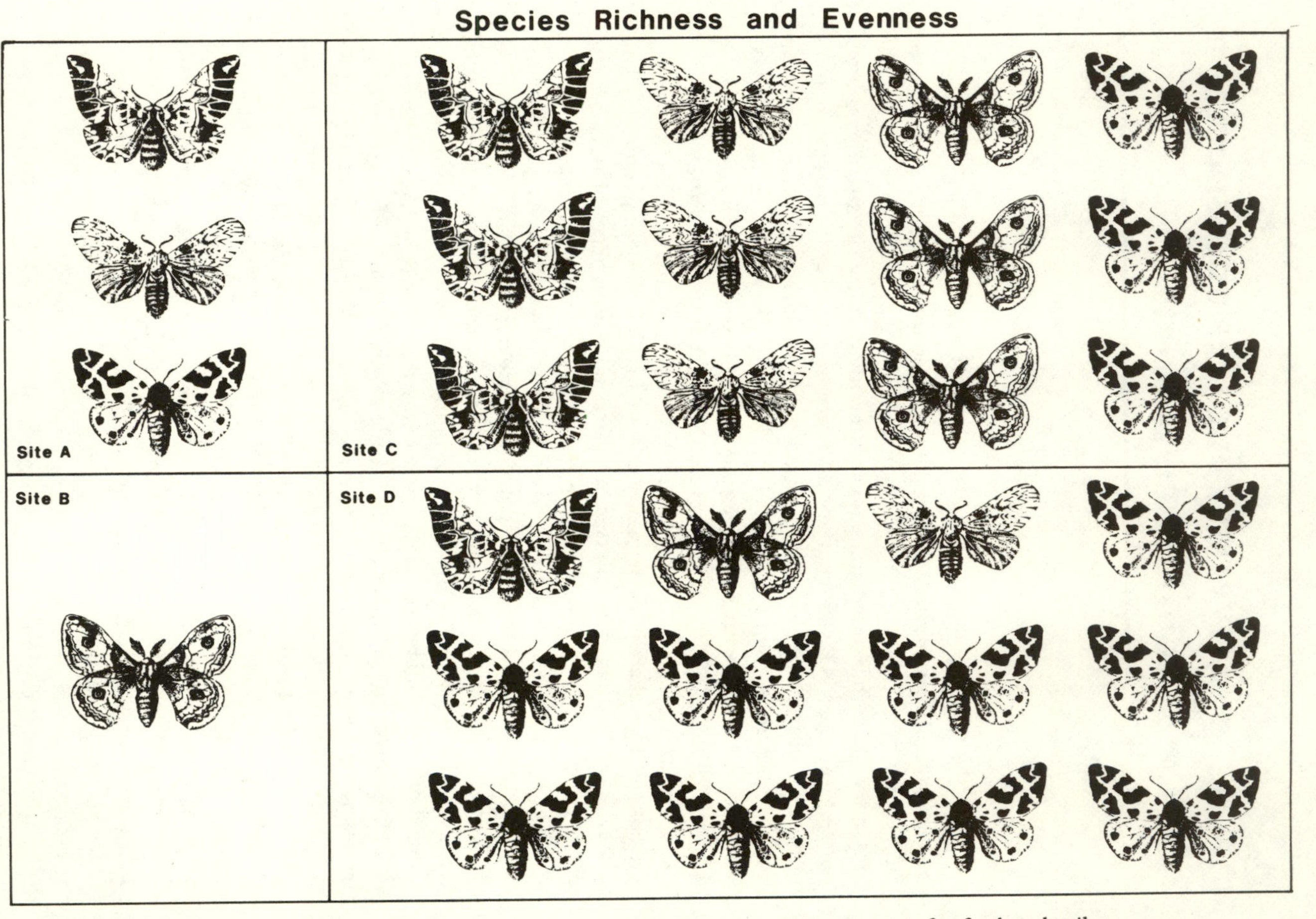

Figure 2.1 A theoretical example to illustrate the concepts of richness and evenness. See text for further details.

number of species in a defined sampling unit. Secondly, there are the *species abundance models* which describe the distribution of species abundances. Species abundance models range from those which represent situations where there is high evenness to those which characterize cases where the abundances of species are very unequal. The diversity of a community may therefore be described by referring to the model which provides the closest fit to the observed pattern of species abundances. If a single diversity index is required a parameter of an appropriate distribution can be used. *Indices based on the proportional abundances of species* form the final group. In this category come the indices such as those of Shannon and Simpson, which seek to crystallize richness and evenness into a single figure.

The remainder of this chapter reviews diversity indices and species abundance models. Sample sizes are considered in Chapter 3 which also discusses the procedures for estimating diversity in situations where, as is the case in many seashore and plant communities, it is difficult to express abundance as numbers of individuals.

Species richness indices

If the study area can be successfully delimited in space and time, and the constituent species enumerated and identified, species richness provides an extremely useful measure of diversity. If however a sample rather than a complete catalogue of species in the community is obtained, it becomes necessary to distinguish between numerical species richness, which is defined as the number of species per specified number of individuals or biomass (Kempton, 1979), and species density, which is the number of species per specified collection area (Hurlbert, 1971). Species density, for example the number of species per m^2, is the most commonly used measure of species richness, and is especially favoured by botanists (see for instance Bunce and Shaw, 1973; Kershaw and Looney, 1985). Numerical species richness on the other hand, although on the whole less frequently adopted, tends to be popular in aquatic studies. Homer (1976), for example, used number of species of fish per 1000 individuals in an investigation of the ecology of an estuarine bay receiving thermal pollution.

It is of course not always possible to ensure that all sample sizes are equal and the number of species invariably increases with sample size and sampling effort (Figures 2.1 and 2.2). To cope with this problem Sanders devised a technique, called Rarefaction, for calculating the number of species expected in each sample if all samples were of a standard size (for example 1000 individuals). Sanders's original formula was subsequently modified by Hurlbert (1971) to produce an unbiased estimate:

$$E(S) = \sum \left\{ 1 - \left[\binom{N-N_i}{n} \Big/ \binom{N}{n} \right] \right\} \quad (2.1)$$

where $E(S)$ = expected number of species;
n = standardized sample size;
N = total number of individuals recorded;
N_i = number of individuals in the ith species.

A worked example is shown in Example 1 (page 127).

A major criticism of rarefaction is that it leads to a great loss of information (Williamson, 1973). This is because the number of species and their relative abundances is known for each sample before rarefaction. After rarefaction all that remains is the expected number of species per sample. Williamson has also criticized Simberloff's (1972) attempt to circumvent the problem by using a computer to select evenly sized samples. A more promising approach is described by Kempton and Wedderburn (1978) who have devised a method for producing equal sized samples from a community in which species abundances are gamma distributed (page 31).

Species richness measures have great intuitive appeal and avoid many of the pitfalls which can be encountered when models and indices are employed. So long as care is taken with sample size (see Chapter 3), species richness measures provide an instantly comprehensible expression of diversity. Species richness, as a measure of diversity, has been used successfully in many studies, for example those of Abbott (1974), Connor and Simberloff (1978) and Harris (1984). However the great range of diversity indices and models which go beyond species richness is evidence of the importance that many ecologists place on information about the relative abundances of species. Kempton (1979)

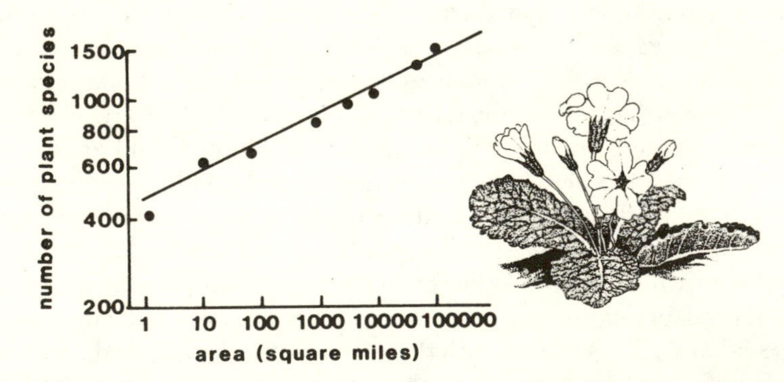

Figure 2.2 Species richness increases with sample size. This graph shows the relationship between number of species and area for flowering plants in England. The smallest sample is of an area of 1 square mile while the largest plot represents the whole of England. Redrawn from Krebs (1985) after Williams (1964).

observes that the distribution of species abundances is often a more sensitive measure of environmental disturbance than species richness alone.

A number of simple indices have been derived using some combination of S (the number of species recorded) and N (the total number of individuals summed over all S species). These include Margalef's diversity index (Clifford and Stephenson, 1975) D_{Mg}

$$D_{Mg} = (S-1)/\ln N \quad (2.2)$$

and Menhinick's index (Whittaker, 1977) D_{Mn}

$$D_{Mn} = S/\sqrt{N} \quad (2.3)$$

[NB: Formulae in this chapter will use natural (i.e. Naperian) logarithms ($\ln = \log_e$) except where explicitly stated otherwise.]

Ease of calculation is one great advantage of Margalef's and Menhinick's indices. For instance, in a sample in which there were 23 species of passerine birds represented by a total of 312 individuals, diversity would be estimated as $D_{Mg} = 3.83$ using Margalef's index and as $D_{Mn} = 1.20$ using Menhinick's index. Convention dictates that Menhinick's index is calculated using S species while Margalef's index uses $S-1$ species. Although it would be more straightforward if both indices were consistent and used either S or $S-1$ it seems best to follow accepted practice and continue to calculate the indices in the usual way. See Example 1 (page 127).

Species abundance models

As data sets containing information on number of species and on their relative abundances were gradually accumulated it was noticed that a characteristic pattern of species abundance was occurring (Fisher *et al.*, 1943). In no community examined would all species be equally common. Instead, as the examples in Figure 2.3 illustrate, it was found that a few species would be very abundant, some would have medium abundance, while most would be represented by only a few individuals. This observation led to the development of species abundance models. These models are strongly advocated by many workers including May (1975, 1981) and Southwood (1978) as providing the only sound basis for the examination of species diversity. A species abundance distribution utilizes all the information gathered in a community and is the most complete mathematical description of the data.

Although species abundance data will frequently be described by one or more of a family of distributions (Pielou, 1975), diversity is usually examined in relation to four main models. These are the log normal distribution, the geometric series, the logarithmic series and MacArthur's broken stick model. When plotted on a rank/abundance graph (Figure 2.4) the four models can be

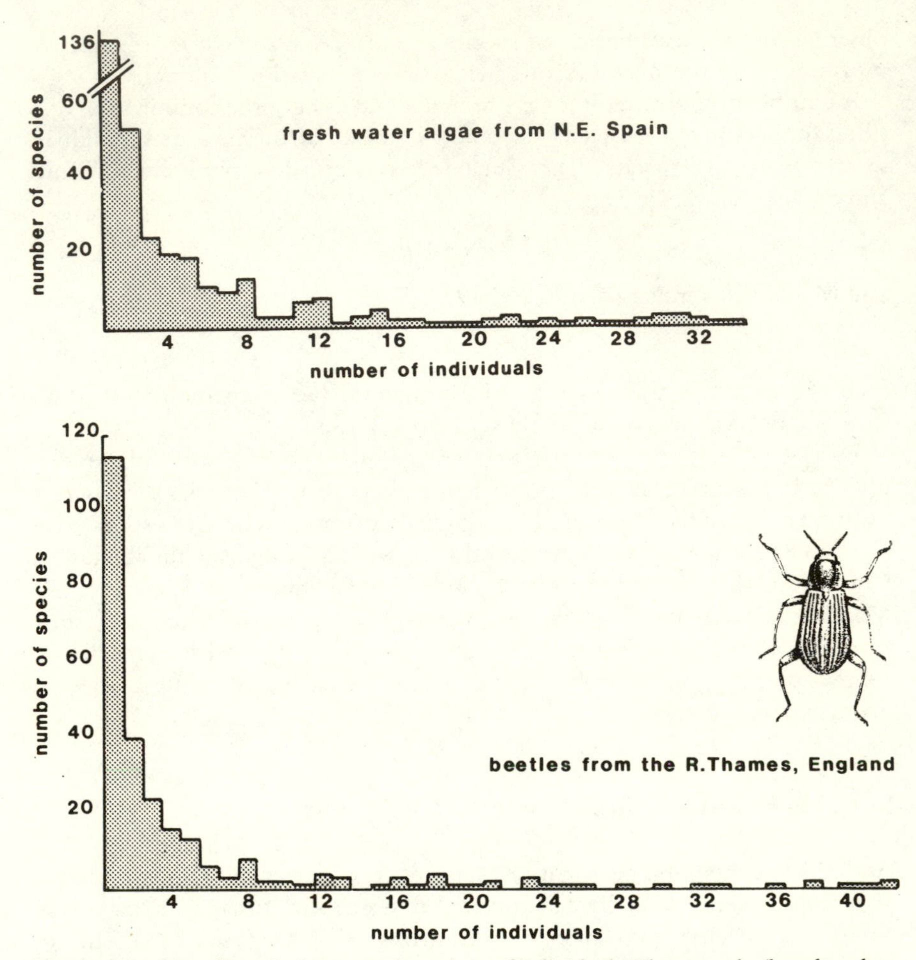

Figure 2.3 Not all species have equal numbers of individuals. These graphs (based on data in Williams, 1964) show the relationship between number of species and number of individuals in two animal communities: fresh-water algae in small ponds in N.E. Spain and beetles in river-flood refuse from the River Thames, England. The majority of species in both cases are represented by only a single individual while a few species in the two samples are very abundant.

seen to represent a progression ranging from the geometric series where a few species are dominant with the remainder fairly uncommon, through the log series and log normal distributions where species of intermediate abundance become more common and ending in the conditions represented by the broken stick model in which species are as equally abundant as is ever observed in the real world.

This arrangement can also be considered in terms of resource partitioning where the abundance of a species is in some way equivalent to the portion of niche space it has pre-empted (or occupied). As Southwood (1978) points out, the geometric series (sometimes called the niche pre-emption hypothesis) represents a situation of maximal niche pre-emption (where a few species dominate, that is they have pre-empted a large proportion of the niche hyperspace), while the broken stick model reflects a case of minimal pre-emption with resources much more equally divided. It is obvious from this discussion that evenness will be high if the broken stick model applies and low if the geometric series is the best fit.

The models each have a characteristic shape on a rank/abundance plot (Figure 2.4) (Whittaker, 1977). The geometric series appears as a straight line with steep gradient. Likewise the log series has a steep gradient but here the curve is only approximately linear. By far the flattest curve is produced by the broken stick model. In between the log series and broken stick comes the log normal with its sigmoid curve. Although this method of plotting is widely used in diversity studies, inspection of a rank/abundance plot is not a failsafe guide to the model that provides the best description of the data. To be certain it is necessary to formally test mathematical fit. The methods of doing this are described below.

Methods of plotting species abundance data

Rank/abundance plots are only one method of presenting species abundance data (May, 1975). They are frequently used by people investigating the geometric series. Proponents of the log series on the other hand often favour a frequency distribution in which number of species is plotted against number of individuals per species (see for example Figure 2.3). A similar plot is used, but with the x-axis on a log scale, when the log normal is chosen (Preston, 1962, and Figures 2.7, 2.8, 2.10 and 2.11). By contrast, when the broken stick model is under investigation a rank/abundance plot, in which the ranks but not abundances are logged, is adopted (Figure 2.5B and King, 1964). These various types of plots highlight the aspect of the data which the ecologist may perhaps wish to emphasize; in the broken stick 'preferred-plot' a straight line, signifying equal abundances, is produced, in the geometric series 'preferred-plot' the few dominant and many rare species are shown, and in the log normal 'preferred-plot' a normal curve, where the eye is drawn to the preponderance of species of intermediate abundance, is obtained.

The range of methods used to display species abundance data has done little to lessen the confusion which besets the measurement of diversity. In 1975 May argued forcibly for a standardization of methods of plotting which would facilitate a more ready comparison of different data sets. Unfortunately, there still seems to be little progress in that direction.

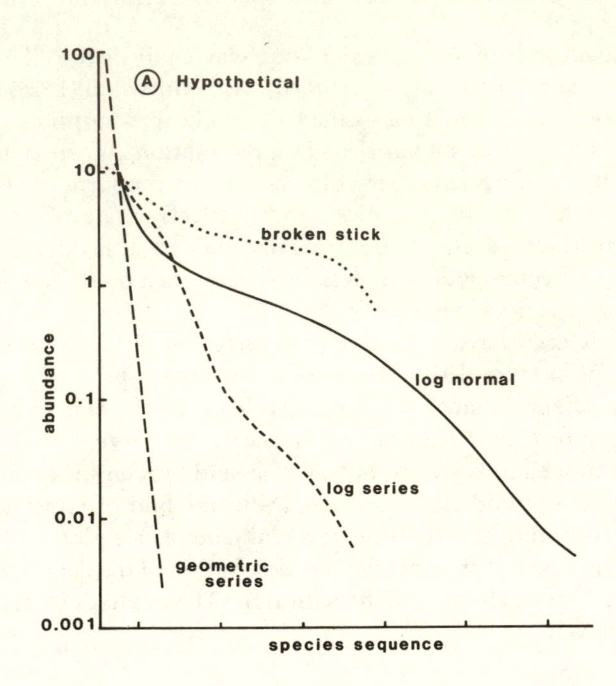

A Hypothetical
100
10
1
0.1
0.01
0.001
abundance
broken stick
log normal
log series
geometric series
species sequence

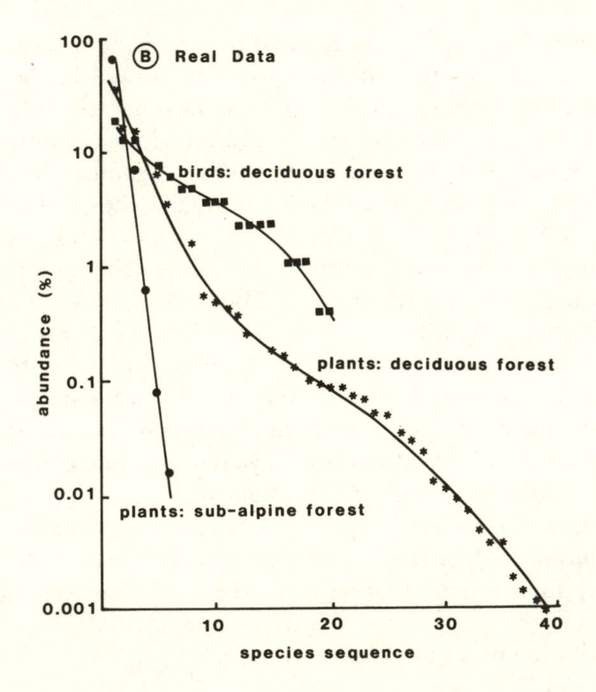

B Real Data
100
10
1
0.1
0.01
0.001
abundance (%)
birds: deciduous forest
plants: deciduous forest
plants: sub-alpine forest
10
20
30
40
species sequence

One recent addition to the catalogue of graphical methods is the k-dominance plot of Platt *et al.* (1984) in which percentage cumulative abundance is plotted against log species rank (Figure 2.5B). The graph obtained is essentially the inverse of the 'broken stick' plot described above. Platt *et al.* (1984) argue that diversity can only be unambiguously assessed when the k-dominance curves from the communities to be compared do not overlap. In this situation the lowest curve will represent the most diverse community. If the curves do intersect Platt *et al.* (1984) claim that it is impossible to discriminate between the communities according to diversity as different diversity indices rank them in opposite ways. This finding merely reflects the observation expanded more fully at the end of this chapter and in Chapter 4 that diversity indices focus on one aspect of the species abundance relationship and emphasize either species richness or dominance. In fact, contrary to the assertion of Platt *et al.* (1984), k-dominance diversity plots which intersect may be the most informative in that they illustrate the shift of dominance relative to that of species richness. This would be similar to the way in which graphs are used to determine the direction of a significant interaction in an analysis of variance (Sokal and Rohlf, 1981). Gray (1988) has also criticized the k-dominance plot as being overly dependent on the abundance of the most abundant species. A diversity measure, the Q statistic, which is based on a cumulative abundance plot, but has the virtue of not relying on information at either end of the curve, is discussed on page 32.

The geometric series

Visualize a situation in which the dominant species pre-empts proportion k of some limiting resource, with the second most dominant species pre-empting the same proportion k of the remainder, the third species taking k of what is left and so on until all species (S) have been accommodated. If this assumption is

Figure 2.4 Rank abundance plots illustrating the typical shape of four species abundance models: geometric series, log series, log normal and broken stick. In these graphs the abundance of each species is plotted on a logarithmic scale against the species' rank, in order from the most abundant to least abundant species. Species abundances may in some instances be expressed as percentages to provide a more direct comparison between communities with different numbers of species. (A) Hypothetical curves to illustrate typical shapes of the four models on a rank abundance plot. (B) Three examples of rank abundance curves from real communities (redrawn from Whittaker, 1970). The three communities are nesting birds in a deciduous forest, West Virginia, vascular plants in a deciduous cove forest in the Great Smoky Mountains, Tennessee, and vascular plant species from sub-alpine fir forest, also in the Great Smoky Mountains. As comparison with (A) suggests, the best descriptions of these three communities are respectively the broken stick, log normal and geometric series models.

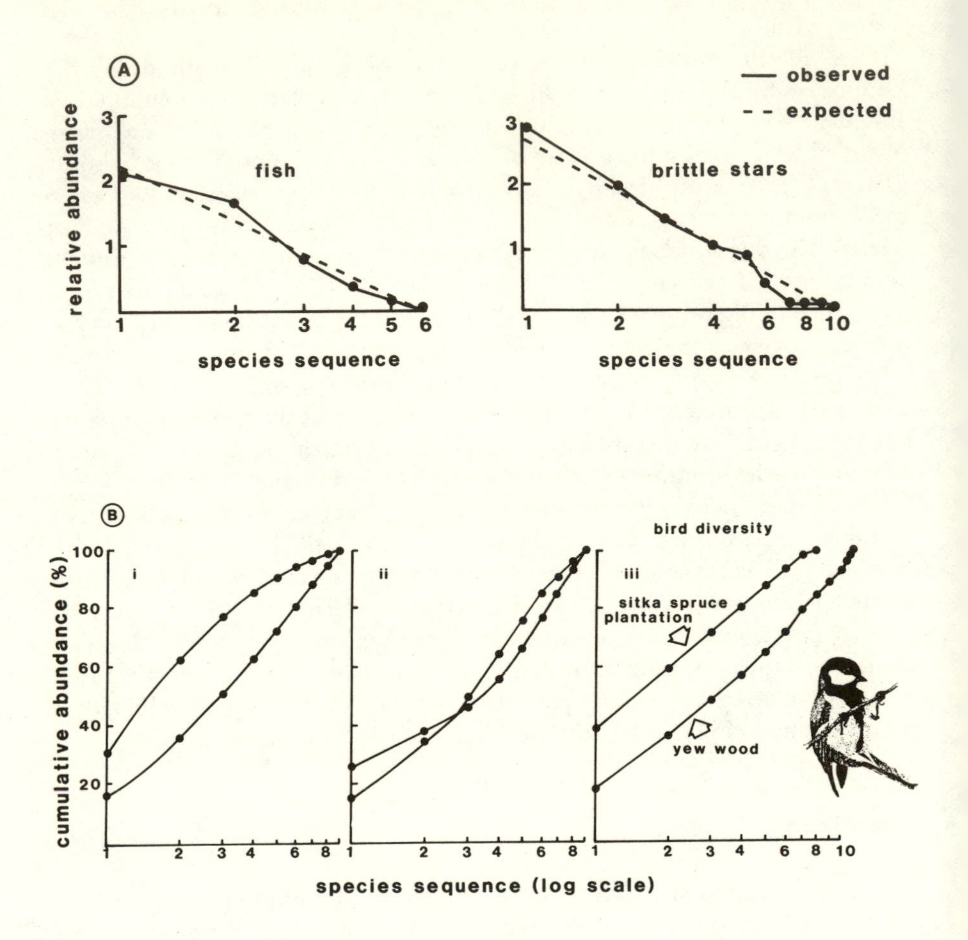

Figure 2.5 Other methods of plotting diversity data. (A) The typical plot used in conjunction with the broken stick model. Relative abundance is plotted in a linear scale on the *y*-axis while the logged species sequences (in order for most abundant to least abundant species) are plotted on the *x*-axis. The two graphs show the observed and expected abundances of fish (family Percidae) and brittle stars (ophiuroids). Redrawn from King (1964). (B) The k-dominance plot in which percentage cumulative abundance is plotted against the log of species rank. Examples i and ii are hypothetical. Platt *et al.* (1984) argue that diversity can only be unambiguously assessed when the k-dominance plots do not overlap (for example in graph i). In this situation the upper curve will be from the more dominant and hence the less diverse assemblage. Where the curves do cross (example ii) it is not possible to rank the communities according to their diversity simply by examining the graph (but see the text for a fuller discussion). Example iii shows k-dominance plots for bird diversity in a sitka spruce plantation and a native yew wood in Killarney, Ireland (data from Batten, 1976). In this comparison the sitka spruce plantation is clearly less diverse.

fulfilled and if the abundances of species (measured for example by biomass or number of individuals) are proportional to the amount of the resource that they utilize, the resulting pattern of species abundances will follow the geometric series (or niche pre-emption hypothesis). In a geometric series the abundances of species ranked from most to least abundant will be (May, 1975; Motomura, 1932):

$$n_i = NC_k k(1-k)^{i-1} \tag{2.4}$$

where n_i = the number of individuals in the ith species;
N = the total number of individuals;
$C_k = [1-(1-k)^s]^{-1}$ and is a constant which ensures that $\Sigma n_1 = N$.

Because the ratio of the abundance of each species to the abundance of its predecessor is constant through the ranked list of species the series will appear as a straight line if plotted on a log abundance/species rank graph (Figure 2.4). Drawing this type of plot is the easiest method of deciding whether a set of data follow the geometric series. Example 2 (page 130) gives some further mathematical details as well as some suggestions about what to do when not all points fall on a straight line. A full mathematical treatment of the geometric series is to be found in May (1975) who has also obtained the species abundance distribution corresponding to the rank abundance series.

Field data have shown that the geometric series pattern of species abundance is found primarily in species-poor (and often harsh) environments or in the very early stages of a succession (Whittaker, 1965, 1970, 1972). As succession proceeds, or as conditions ameliorate, species abundance patterns grade into those of the log series.

The log series

Fisher's logarithmic series model (Fisher *et al.*, 1943) represented the first attempt to describe mathematically the relationship between the number of species and the number of individuals in those species. Although originally used as an appropriate fit to empirical data, its wide application, especially in entomological research, has led to a thorough examination of its properties (Taylor, 1978). Many authors, including Southwood (1978), make a distinction between the log series and the geometric series, but, as May (1975) notes, the geometric series and log series models are closely related. For instance Thomas and Shattock (1986) found that both the geometric and log series adequately described the species abundance pattern of filamentous fungi on the grass *Lolium perenne*. The geometric series would be predicted to occur in a situation in which species arrived at an unsaturated habitat at regular intervals of time, and occupied fractions of remaining niche hyperspace. A log series pattern would however result if the intervals between the arrival of these species were random rather than regular (Boswell and Patil, 1971; May, 1975). The small number of abundant species and the large proportion of 'rare' species (the class containing one individual is always the largest) predicted by the log series model suggest that, like the geometric series, it will be most applicable in situations where one or a few factors dominate the ecology of a community. For instance Magurran (1981) showed that species abundances of ground flora in an Irish conifer plantation (in which light is greatly limited) followed a log series distribution (Figure 2.6 and see Chapter 4).

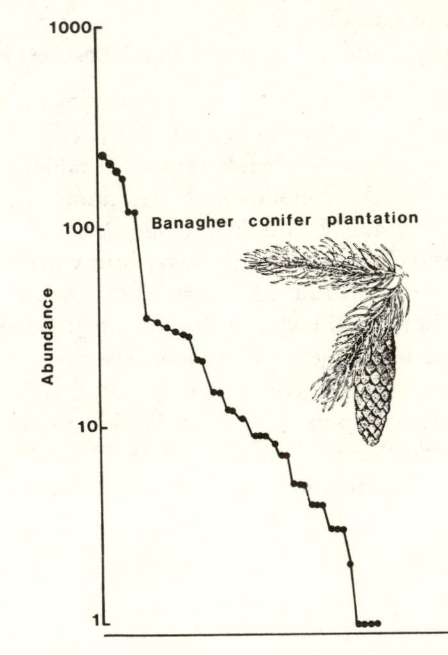

Figure 2.6 A rank abundance plot showing the diversity of ground vegetation in an Irish conifer plantation (for more information on the sites see Figure 4.2 and Chapter 4). One factor, light, has an important influence on the diversity of the vegetation, and species abundances follow a log series distribution. For a comparison with the diversity of ground vegetation in an adjacent natural deciduous woodland, see Figures 2.7 and 4.3.

It should be noted that, when sample sizes are small, the log series may arise as a sampling distribution (May, 1975 and see below under log normal).

The log series takes the form:

$$\alpha x, \frac{\alpha x^2}{2}, \frac{\alpha x^3}{3}, \ldots \frac{\alpha x^n}{n} \tag{2.5}$$

αx being the number of species predicted to have one individual, $\alpha x^2/2$ those with two and so on (Fisher *et al.*, 1943; Poole, 1974).

The total number of species, S, is obtained by adding all the terms in the series which reduces to the following equation

$$S=\alpha[-\ln(1-x)] \tag{2.6}$$

x is estimated from the iterative solution of

$$S/N=(1-x)/x[-\ln(1-x)] \tag{2.7}$$

where $N=$ the total number of individuals.

In practice x is almost always >0.9 and never >1.0. If the ratio $N/S>20$ then $x>0.99$ (Poole, 1974).

Two parameters, α, the log series index, and N, summarize the distribution completely, and are related by

$$N=\alpha\ln(1+N/\alpha) \tag{2.8}$$

α is an index of diversity. It has been widely used, and remains popular (Taylor, 1978), despite the vagaries of index fashion.

The index may be obtained from the equation

$$\alpha=\frac{N(1-x)}{x} \tag{2.9}$$

with confidence limits set by

$$\text{Var}(\alpha)=\frac{\alpha}{-\ln(1-x)} \tag{2.10}$$

(Taylor *et al.*, 1976) or alternatively α may be read from Williams's nomograph (Williams, 1964).

The procedure for fitting the model is to calculate the number of species expected in each abundance class and compare that with the number of species actually observed using a goodness of fit test (χ^2 or G test; Sokal and Rohlf, 1981). A worked example is shown in Example 3 (page 132). Further mathematical details about the log series are provided by May (1975). A series of studies (Taylor, 1978; Kempton and Taylor, 1974, 1976) investigating the properties of the log series index α have come out strongly in favour of its use, even when the log series distribution is not the best descriptor of the underlying species abundance pattern. The advantages of α and the log series distribution relative to the other models and indices are reviewed in Chapter 4. Chapter 4 also discusses the validity of using goodness of fit tests to decide which model is most appropriate to a particular data set.

Log normal distribution

The majority of communities studied by ecologists display a log normal pattern of species abundance (Sugihara, 1980). Although the log normal model may be said to indicate a large, mature and varied natural community its applicability to other large data sets has been demonstrated. May (1975) for instance has shown that the world distribution of human populations and the distribution of wealth within the USA are both log normal. [In Britain by contrast the pattern of wealth pertains more to the log series, a substantially less equitable state of affairs! (May, 1974).] One explanation for the ubiquity of the log normal stems simply from the mathematics of the distribution. The log normal distribution will arise as a response to the statistical properties of large numbers and as a consequence of the Central Limit Theorem (May, 1975). The Central Limit Theorem states that when a large number of factors act to determine the amount of a variable, random variation in those factors will result in that variable being normally distributed. This effect becomes more true as the number of determining factors increases. In the case of log normal distributions of species abundance data the variable is the number of individuals per species (standardized by a log transformation) and the determining factors all the processes which govern community ecology.

The log normal distribution was first applied to species abundance data by Preston in 1948. Preston plotted species abundances using $\log_2$ and termed the resulting classes octaves. These octaves represent doublings in species abundances (see for example Figure 2.7A). It is not however necessary to use $\log_2$: any log base is valid and $\log_3$ (Figure 2.7B) and $\log_{10}$ (Figure 2.7C) are two common alternatives. May (1975) provides a thorough and lucid discussion of the model.

The distribution is usually written in the form:

$$S(R) = S_0 \exp(-a^2 R^2) \tag{2.11}$$

where $S(R)$ = the number of species in the Rth octave (i.e. class) to the right and left of the symmetrical curve;
S_0 = the number of species in the modal octave;
$a = (2\sigma^2)^{1/2}$ = the inverse width of the distribution.

Empirical studies have shown that a is usually ≈ 0.2 (May, 1981; Whittaker, 1972). One further parameter of the log normal (γ) is also conventionally defined. Like a its value is remarkably consistent across data sets.

γ is illustrated in Figure 2.8. When a curve of the total number of individuals in each octave (the individuals curve) is superimposed on the species curve of the log normal, γ is a measure of the relationship between the mode of the individuals curve and the upper limit of the species curve. Explicitly it is an estimate of the number of species at the octave where the individuals curve reaches its crest.

$$\gamma = R_N / R_{max} = \ln 2 / [2a(\ln S_0)^{1/2}] \quad \text{(May, 1975)} \tag{2.12}$$

where R_N = the modal octave of the individuals curve;

R_{max} = the octave in the species curve containing the most abundant species.

In many cases the crest of the individuals curve (R_N) coincides with the upper tail of the species curve (R_{max}) to give $\gamma \approx 1$. In such log normals, described by Preston (1962) as canonical (Preston's canonical hypothesis) the standard deviation is constrained between narrow limits (giving $a \approx 0.2$). May (1975) showed that the relationship of $\gamma \approx 1$ is also found in log normal distributions of non-ecological data including those of wealth and population mentioned above. He went on to argue that the relationship has no biological basis and is simply an artifact of the mathematical properties of the log normal distribution. Sugihara (1980) however demonstrated that natural communities (including those of birds, moths, gastropods, plants and diatoms) fit the canonical hypothesis too well for this to be the case (Figure 2.9). Species-rich

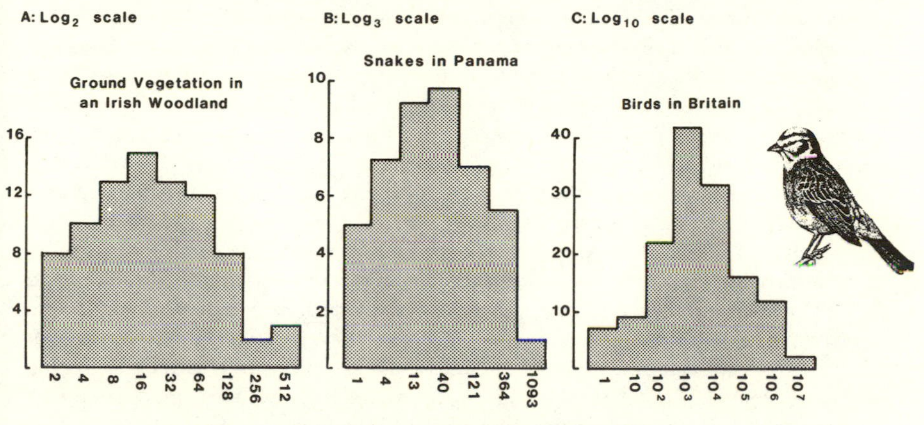

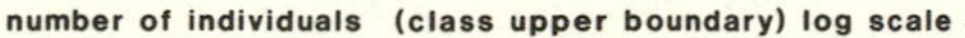

Figure 2.7 The log normal distribution I. The 'normal', symmetrical bell-shaped curve is achieved by logging the species abundances on the *x*-axis. A variety of log bases can be used. (A) $\log_2$. This usage follows Preston (1948). Species abundances are expressed in terms of doublings of numbers of individuals. For example successive classes would be 2 or fewer individuals, 3–4 individuals, 5–8 individuals, 9–16 individuals, 17–32 individuals and so on. It is conventional to call the classes, octaves. The graph shows the diversity of ground vegetation in a natural deciduous woodland at Banagher in N. Ireland (see Figure 4.2 and Chapter 4). (B) $\log_3$. Instead of doublings the successive classes refer to treblings of numbers of individuals. Thus in this example showing the diversity of snakes in Panama (data from Williams, 1964) the upper bounds of the classes are 1, 4, 13, 40, 121, 364 and 1093 individuals. Although used widely by Williams (1964) $\log_3$ is rarely employed today. (C) $\log_{10}$. Classes in $\log_{10}$ represent increases in order of magnitude 1, 10, 100, 1000, 10 000, 100 000. This choice of log base is most appropriate for very large data sets, as for example in this case the diversity of birds in Britain (data from Williams, 1964). In all cases the *y*-axis shows the number of species per class.

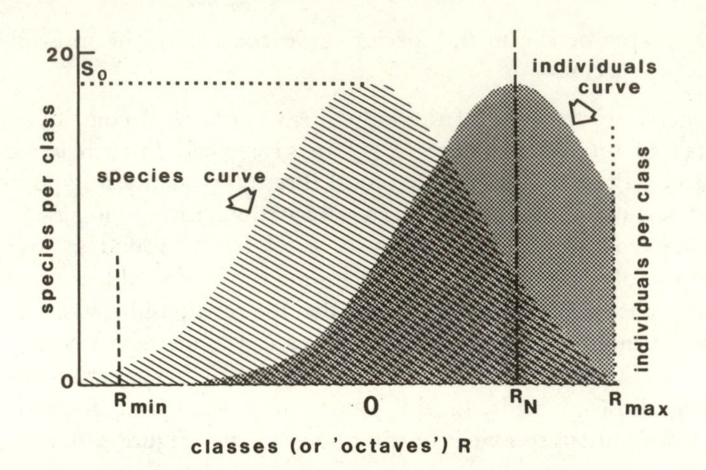

Figure 2.8 The features of the log normal distribution, II. The hatched curve (species curve) shows the distribution of numbers of species amongst classes. (For historical reasons the abundances that these classes represent are often expressed in $\log_2$, or doublings of number of individuals – see Figure 2.7). Since the distribution is symmetrical, classes in the same position on either side of the mode are expected to have equal numbers of species. For this reason it is conventional to term the modal class 0 and refer to classes to the right of the mode as 1, 2, 3, etc. and those to the left hand side of the mode as -1, -2, -3, etc. R_{min} marks the expected position of the least abundant species while R_{max} shows the expected position of the most abundant species and $R_{max} = -R_{min}$. For instance if there were five classes either side of the mode R_{max} would be 5 with R_{min} as -5. The number of species in each class is $S(R)$. Thus in this example the number of species in the modal class, S_0, would be 18. In addition to the species curve, there is an individuals curve which gives the total number of individuals present in each class. The class which contains the most individuals (that is the one in which the mode of the individuals curve occurs) is termed R_N. A log normal distribution is described as canonical when R_N and R_{max} coincide to give the value of $\gamma = 1$ (where $\gamma = R_N/R_{max}$). Redrawn from May (1975).

communities, that is those with 200 or more species, are most likely to be canonical (Ugland and Gray, 1982).

Sugihara (1980) has proposed a biological explanation for the canonical log normal distribution of species abundances. He envisages the communal (multidimensional) niche space of a taxon being sequentially split by the constituent species. The portion of niche space each species occupies is proportional to its relative abundance and the probability of any fragment of niche being subdivided is independent of its size. Sugihara has likened the process to a rock crushing operation (where the sizes of the resulting pieces of gravel will be log normally distributed). Such a process could arise either through an ecological or an evolutionary mechanism.

There are an infinite number of ways in which resources can be split using Sugihara's model and other methods of division will yield different species

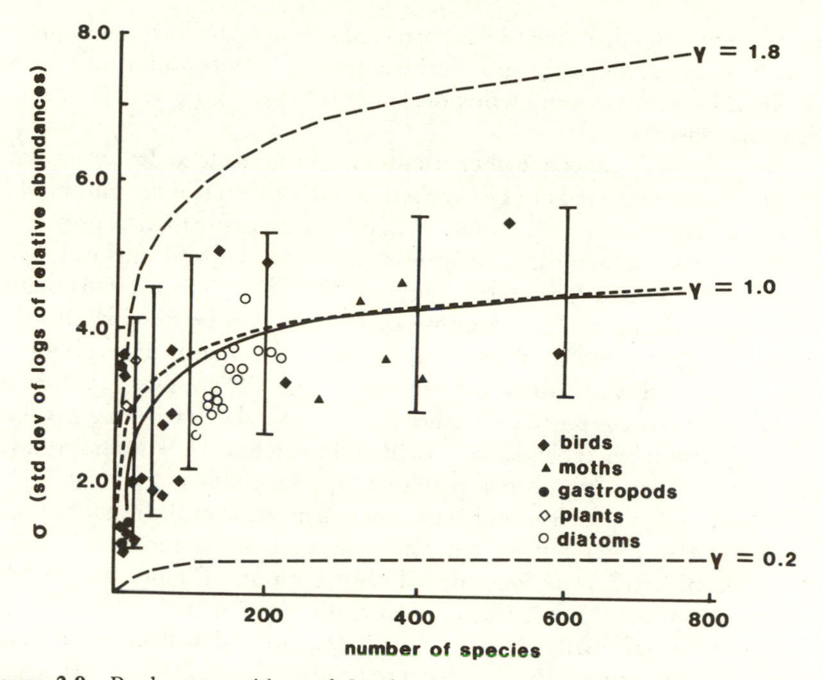

Figure 2.9 Real communities and Sugihara's sequential niche breakage model. This figure (redrawn from May, 1981, after Sugihara, 1980) shows the relationship between species richness, S, and the standard deviation, σ, of the logged relative abundances. The three dashed lines illustrate the form of the relationship for log normal distributions in which $\gamma = 1.8$, $\gamma = 1.0$ (canonical log normal) and $\gamma = 0.2$, while the solid line represents Sugihara's prediction (with error bars showing two standard deviations either side of the mean). Sugihara's model shows a close agreement with the canonical log normal. In addition the real communities of birds, moths, gastropods, plants and diatoms cluster tightly around the line representing the canonical log normal.

abundance distributions. For instance, if the smallest portion of niche space is always the one which is split, a log series will result. Splitting the largest portion will produce a very equitable distribution.

Two factors distinguish Sugihara's sequential breakage hypothesis from other resource partitioning models. First, unlike the broken stick (see below) and geometric series, niche space in Sugihara's model can be multidimensional. Secondly, it requires that the breakages take place successively. In the broken stick model the breakages are simultaneous.

One model which is similar to Sugihara's is Pielou's (1975) sequential breakage model. This restricts itself to one resource axis which is randomly and sequentially split. Though a log normal distribution results, Pielou does not specify whether it is canonical.

As May (1981) emphasizes, the correlation with empirical data is no guarantee that Sugihara's model is correct. The model however does provide us with an excellent working hypothesis for the diversification of niches in ecological communities and is flexible enough to generate a variety of species abundance distributions.

Since Sugihara's paper a further attempt has been made to demonstrate that ecological processes need not be invoked in order to explain the canonical log normal. Ugland and Gray (1982) show that $\gamma \approx 1$ is a mathematical property of log normal distributions based on 50 or more species (Ugland and Gray, 1982). Ugland and Gray (1982) have also suggested why the log normal is so common in ecological data sets. They propose that species can be divided into three classes: rare species (65% of the total), species with intermediate population sizes (25%) and very abundant species (10%). Then they assume that communities are composed of patches and that the abundance of a particular species is the sum of its abundance in each of the patches. These assumptions are sufficient to create a log normal pattern of species abundance.

Speculation has also surrounded the consistent value of the other canonical parameter a ($a \approx 0.2$) but to date it appears that the result is simply a mathematical artifact of log normal distributions of moderate or large numbers of species (May, 1975; Ugland and Gray, 1982).

The log normal distribution is a symmetrical 'normal' bell-shaped curve. If, however, the data to which the curve is to be fitted derive from a finite sample, the left hand portion of the curve (representing the rare and consequently unsampled species) will be obscured. Preston (1948) terms the truncation point of the curve the veil line, and, the smaller the sample, the further this veil line will be from the origin of the curve (Figure 2.10). In most data sets only the portion of the curve to the right of the mode will be visible and it is only in immense data collections covering wide biogeographic areas that the full curve is apparent (Figure 2.11).

Fitting the log normal would be simple if it were not for the problem of the veil line. Pielou (1975) has however devised a method for fitting a truncated log normal. This method makes the assumption that the position of the veil

---

Figure 2.10 (A) The veil line is illustrated in this figure (redrawn from Taylor, 1978). In small samples only the portion of the distribution to the right of the mode is apparent. However as sample size increases the veil line moves to the left, revealing first the mode and eventually the entire log normal distribution. This effect is shown in (B). (B) Fish diversity in the Arabian Gulf. Samples of fish were collected in an area of the Gulf adjacent to Bahrain. Abundance is expressed as the mean number of individuals caught in 45 min trawling and is plotted on the x-axis using log $base_2$. In single samples (for example one taken in May) only the right hand portion of the log normal distribution is evident. By taking all the samples from May and June together it is possible to see the mode, and with an entire year's data the full log normal distribution is revealed (Magurran and Abdulquadar, unpublished data). A similar effect can be seen in Figure 2.13.

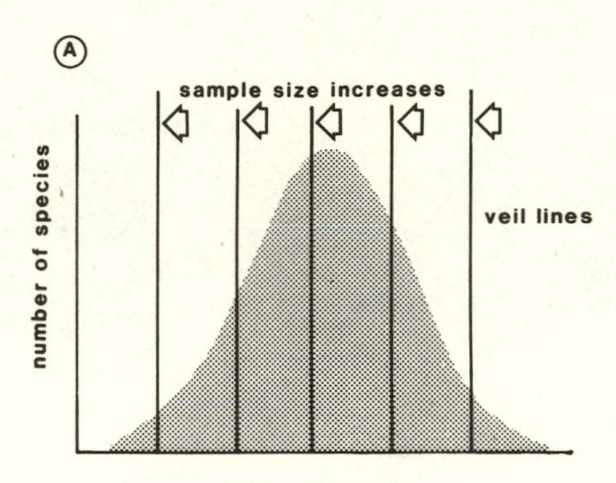

A
sample size increases
veil lines
number of species
log number of individuals

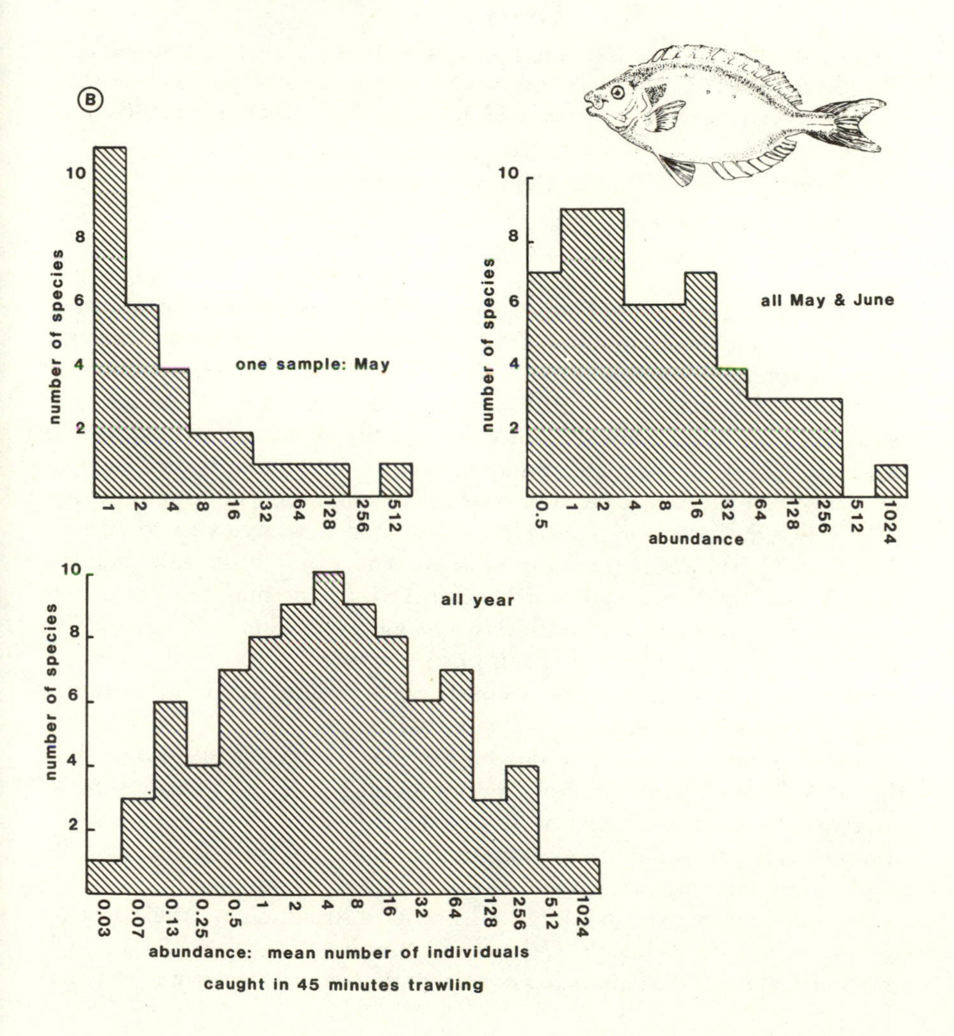

B
number of species
10
8
6
4
2
one sample: May
1
2
4
8
16
32
64
128
256
512
all May & June
0.5
1024
abundance
all year
0.03
0.07
0.13
0.25
abundance: mean number of individuals
caught in 45 minutes trawling

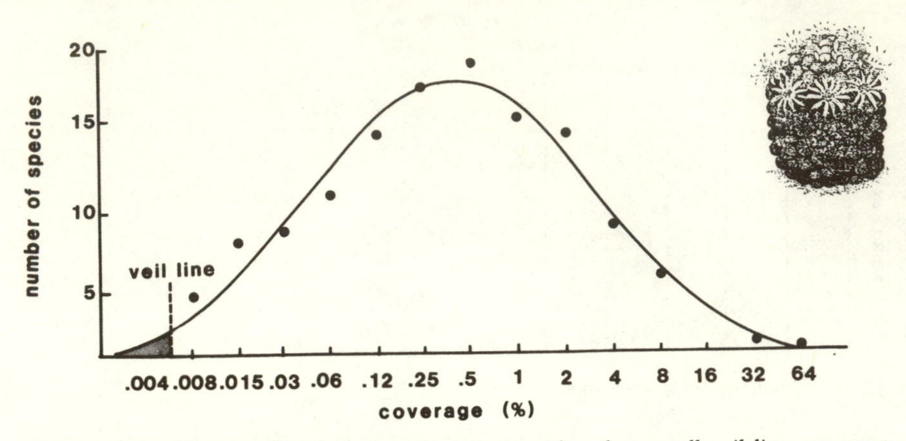

Figure 2.11 The complete log normal, or ones with only a small veil line, are most evident in large data sets. This figure (redrawn from Whittaker, 1965) shows a log normal distribution of plant species abundancies in a Sonoran semidesert. The equation for the fitted curve is:

$$S(R) = 17.5 \exp(-0.245^2 R^2)$$

where $S_0 = 17.5$ and $a = 0.245$.

line or truncation point can be recognized. The procedure entails converting the observed variate (the number of individuals per species) to logs and fitting a normal curve, disregarding the area to the left of the truncation point. The truncation point falls at -0.30103 or $\log_{10} 0.5$, this being the lower class boundary of the class containing those species for which one individual was observed. The area under the remaining part of the curve is then used to estimate S^*, the total number of species in the community. Example 4 (page 136) shows the calculations. In Pielou's method it is necessary to consult Table 1 in Cohen (1961) (reproduced in Appendix 3) in order to obtain the value θ (the auxiliary estimation function) which permits the mean and variance of the truncated distribution to be estimated. Slocomb *et al.* (1977) have automated this process in a computer program.

Pielou's method can now be criticized as being a little dated. It is however retained in this book because it is easy to use.

Strictly speaking the continuous log normal (whether truncated or not) should be fitted only to continuous species abundance data such as measures of cover or biomass (see Chapter 3). In practice, however, most people use the continuous log normal when working with numbers of individuals, since, for large sample sizes especially, the data are effectively continuous.

An alternative method of fitting a log normal distribution to sample data has been discussed by Bulmer (1974) and Kempton and Taylor (1974) and is referred to as either the Poisson log normal or the discrete log normal. Here it is assumed that the continuous log normal curve is represented by a series of discrete species abundance classes which behave as compound Poisson variates. The Poisson parameter λ is distributed log normally. In practice the Poisson log normal presents greater computational difficulties than the truncated log normal. The values of S^* which it gives have been shown to differ from estimates of S^* produced by the continuous log normal. However it is possible to calculate the variance of S^* for the Poisson log normal while the variance of S^* for Pielou's truncated log normal is as yet unknown. Estimates of S^* derived from the truncated log normal should be treated with extreme caution as the results in Figure 2.12 show.

It might be expected that when a log normal had been fitted σ (the standard deviation) would provide a useful measure of diversity. Although σ gives a measure of evenness (equitability) it is a poor index for discriminating between samples, and cannot be estimated accurately when sample size is small (Kempton and Taylor, 1974). These criticisms do not however apply to the ratio S^*/σ, referred to as λ. There is a marked correlation between the values of λ and α calculated for the same data and both have been shown to provide an

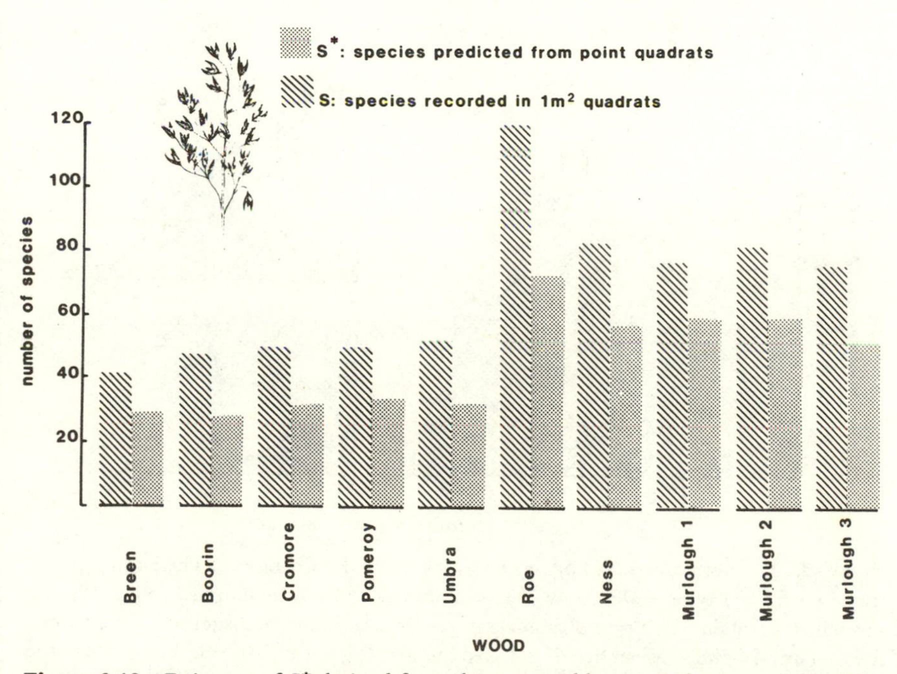

Figure 2.12 Estimates of S^* derived from the truncated log normal are unreliable. This graph shows the discrepancy between the number of species (S) recorded in 50 m^2 quadrats in ten woodlands and the number of species estimated (S^*) from 50 point quadrats placed in the centre of the same quadrats. For a map of the ten woodlands see Figure 6.6.

efficient means of discriminating between samples (Chapter 4; Taylor, 1978; Kempton and Taylor, 1974).

Many data sets will be described equally well by both the log series and the log normal models and it may be difficult for the ecologist to decide which is more appropriate. Figure 2.13 shows that when it is in its truncated form the log normal is virtually indistinguishable from the log series. May (1975) prefers the log normal distribution as he argues that it reflects the many processes at work in a community's ecology. The log normal also describes more data sets than the log series making it a more suitable vehicle for comparing communities. Taylor (1978) and Kempton and Wedderburn (1978) on the other hand favour the log series because it is a poorer fit at the 'rare' end of the curve, especially in large data sets. They feel that this property will ensure that only the resident population in a habitat are considered. Vagrant species will be ignored.

Lambshead and Platt (1985) and Hughes (1986) have recently challenged the

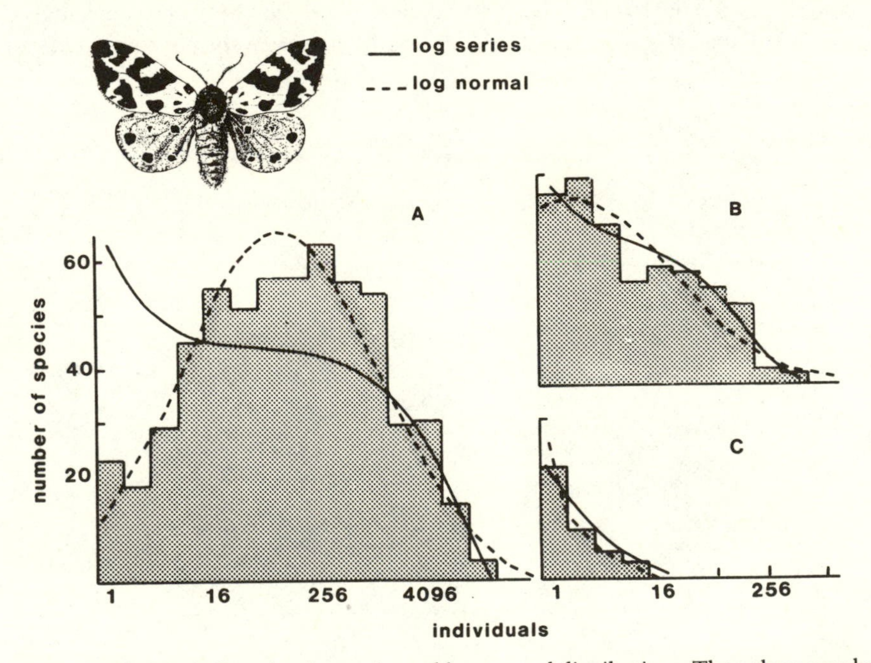

Figure 2.13 Moth diversity. Log series and log normal distributions. These three graphs (redrawn from Taylor, 1978) show (A) the abundance of moths summed across 225 sites throughout Britain, (B) a typical annual sample from a single rural site, and (C) a sample from an impoverished urban site. The dotted lines are log normal distributions fitted to the data. Log series distributions are indicated by solid lines. These graphs demonstrate that small samples (in which the full log normal distribution is veiled) are described equally well by both the log series and (truncated) log normal models. When the complete distribution is revealed the log series ceases to be a good fit.

assertion that most communities are log normal. Lambshead and Platt argue that many classic data sets are not true samples but are in fact collections or amalgamations of non-replicate samples. Furthermore they assert that the shape of the log normal distribution of species abundance is independent of sample size and that there is no evidence of the veil line moving to the left as sample size is increased. They conclude that 'the log normal . . . is never found in genuine ecological samples' and as a consequence they feel that the log series should be adopted when species abundance data are being investigated. Hughes (1986) suggests that the mode which distinguishes the log normal from the log series model may arise from species misidentification and sampling errors. He also feels that the reduction of species abundance classes achieved by use of $\log_3$ or $\log_{10}$ may generate a mode which would not be apparent if the data had been plotted in classes using $\log_2$. This latter criticism may have relevance where small data sets are concerned but is unlikely to affect large sample sizes seriously. Hughes does not favour the log series in place of the log normal; instead he advocates his own dynamics model (see page 31) which he claims is much more widely applicable than either of the two 'traditional' models. While Hughes and Lamshead and Platt rightly draw our attention to the inadequacies of many classic data sets and prove that (as we might expect) the log normal will result if samples are indiscriminately combined, there are still many cases of rigorous sampling yielding genuine log normal distributions (Taylor, 1978; Sugihara, 1980). Thus it seems likely that the log normal will remain an important tool in diversity studies.

The broken stick model

The broken stick model (sometimes called the random niche boundary hypothesis) was proposed by MacArthur in 1957. He likened the subdivision of niche space within a community to a stick broken randomly and simultaneously into S pieces. Unlike Sugihara's log normal model the broken stick is concerned with just one resource. The broken stick model reflects a much more equitable state of affairs than those suggested by the log normal, log series and geometric series. It is the biologically realistic expression of a uniform distribution. A major criticism of the model is that it may be derived from more than one set of hypotheses (Pielou, 1975), and, that as it is characterized by only one parameter, S (number of species), it is strongly subject to sample size (Cohen, 1968; Poole, 1974). Nevertheless, if a broken stick distribution is observed we have evidence that an important ecological factor is being shared more or less evenly between the species (May, 1974). The criticism of being derived from more than one hypothesis can of course be directed at other species abundance distributions as well.

Like the geometric series the broken stick distribution is conventionally written in terms of rank order abundance and the number of individuals in the ith most abundant of S species (N_i) is obtained from the term (May, 1975):

$$N_i = N/S \sum_{n=i}^{S} 1/n \qquad (2.13)$$

where N = total number of individuals;
S = total number of species.

May (1975), after Webb (1974), expresses the model in terms of a standard species abundance distribution.

$$S(n) = [S(S-1)/N]\,(1-n/N)^{S-2} \qquad (2.14)$$

where $S(n)$ = the number of species in the abundance class with n individuals.

As with the log series and log normal distributions discussed above, a goodness of fit test is used to compare the observed and expected frequencies in abundance classes. Example 5 (page 139) shows how this is done. Strictly speaking the broken stick predicts the average species abundance distribution for a number of communities and it can therefore be misleading to test its fit in relation to a single sample or community (Pielou, 1975). However this criticism only applies if it is desired to test the model in the context of MacArthur's precise portrayal of resource partitioning. It is perfectly valid to use the broken stick model as a means of saying that the species abundances in a particular community are more even than would have been the case if the log series, or even the log normal, had produced the best fit.

The broken stick model has been used successfully in a few studies, for example passerine birds (MacArthur, 1960), minnows and gastropods (King, 1964). Good fits of the model seem to be found primarily in narrowly defined communities of taxonomically related organisms.

No diversity index has been derived from the distribution: since it represents a highly equitable state of affairs S (species richness) is an adequate measure of diversity.

MacArthur (1957) also proposed the overlapping niche model which reflects an even greater degree of evenness than is embodied in the broken stick model. Although ecologically unrealistic (Pielou and Arnason, 1965), Pielou (1975) argues that the model should not be rejected out of hand and shows how it can be applied to the analysis of zone widths along an environmental gradient.

The continuum of dominance to evenness terminates with the uniform distribution, in which all species are equally abundant. This distribution exists nowhere in nature though it may sometimes be found in the minds of ecologists who wish to test the performance of various indices and models.

Other distributions

Life would be reasonably simple if there were only four species abundance distributions to contend with. However dissatisfaction with existing models has prompted ecologists to widen their scope. One recent acquisition is the Zipf–Mandelbrot model (Zipf, 1965; Mandelbrot, 1977; Gray, 1987), which, like the Shannon index, has its roots in linguistics and information theory. In an ecological setting the Zipf–Mandelbrot model is interpreted as reflecting a successional process in which later colonists have more specific requirements and hence are rarer than the first species to arrive (Frontier, 1985). The model also postulates a rigid sequence of colonists, with the same species always present at the same point of successions in similar habitats. This prediction is patently not followed in the real world (Gray, 1987) Although the model is usually a poor fit where rare species are concerned it has been successfully applied to a number of communities (Reichelt and Bradbury, 1984; Frontier, 1985; Gray, 1988).

Another recent recruit to the diversity model club is the dynamics model of Hughes (1984, 1986). Hughes developed the dynamics model to explain the patterns of species abundance which characteristically arise in marine benthic communities. In these assemblages there are more abundant species than would be predicted by the log series model yet too few rare species to produce the mode found in log normal distributions. By visually inspecting the rank abundance plots from 222 plant and animal communities Hughes concluded that his dynamics model gave a much better prediction of species abundance pattern than either the log normal or log series models.

One final model which has attracted the attention of diversity students is the truncated negative binomial distribution (Pielou, 1975). Ecologists are most familiar with the negative binomial in its single species application where it is used to distinguish clumped populations from randomly or evenly dispersed ones (Southwood, 1978). Pielou (1975) shows how the distribution can be applied to species abundance data. She also makes the point that if abundances are measured on a continuous scale, for example as biomass or cover, rather than on the discrete scale of number of individuals, it is appropriate to use the gamma distribution rather than the negative binomial.

The negative binomial is mathematically related to both the Poisson series and the log series (Southwood, 1978). The clumping parameter k, which is usually around 2 for the negative binomial, reduces to zero for the log series. If k is infinity the distribution is identical with the Poisson. Southwood (1978) gives more mathematical details.

Although there are some instances where the truncated negative binomial, gamma, dynamics and Zipf–Mandelbrot distributions are good descriptors of ecological data, it would seem prudent to use the four conventional models (geometric series, log series, log normal and broken stick) wherever possible.

This procedure may not provide the intellectual excitement of searching out even more models to test in relation to species abundance data, but at least it should make assembled data sets easier to compare. While it is possible that someone may come up with a model which will revolutionize our understanding of species abundance relationships, at present it seems best to agree with Gray (1988) who concludes that 'the search for yet more models is unlikely to give any insights into factors structuring biological assemblages'.

Biological versus statistical models

The species abundance distributions described above have been loosely classified in two ways. First the models were arranged on a dominance-evenness scale, starting with the geometric series and concluding with the broken stick. Next the less frequently applied models, for instance the gamma, were distinguished from the mainstream ones such as the log normal. A third way of classifying distributions is on the basis of whether they are biological or resource-apportioning (do they make any specific predictions about the ecological processes needed to generate a specific pattern of species abundance?) or statistical (in other words nothing more than a mathematical fit to empirical data). Unfortunately the dichotomy is not clear cut. Only three, the geometric series, the overlapping niche model and the broken stick have a biological pedigree (Pielou, 1975; Gray, 1988). Of these the overlapping niche model is rarely used, and since it does not imply competition for a limited resource should strictly speaking be removed to the statistical camp (Pielou, 1975). The geometric series is restricted in its application and the ecological assumptions of the broken stick are discredited. Pure statistical models include the negative binomial and the log series. The remaining hybrid models were initially statistical but acquired one (for example the Zipf–Mandelbrot) or more (for example the log normal) biological explanations.

To reiterate the point made earlier, there is no reason why a good fit by a particular model vindicates the ecological assumptions that it is based upon. Harvey and Godfray (1987) and Harvey and Lawton (1986) have for instance shown that a canonical log normal distribution of individuals amongst species does not necessarily lead to a canonical log normal distribution of energy utilization. This is because large-bodied species usually have larger energy requirements, but lower population densities, than small-bodied species. Further evidence on the differential resource requirements of large and small-bodied animal has been supplied by Brown and Maurer (1986).

The Q statistic

An interesting approach to the measurement of diversity which takes into account the distribution of species abundances but does not actually entail fitting a model is the Q statistic, proposed by Kempton and Taylor (1976, 1978). This index is a measure of the inter-quartile slope of the cumulative species abundance curve (Figure 2.14) and provides an indication of the diversity of the community, with no weighting either towards very abundant or very rare species. An earlier index suggested by Whittaker (1972) was based on a similar idea. Whittaker's index however considered the full species abundance curve and was subject to bias at both ends of the distribution.

Estimated from empirical data:

$$Q = \frac{\frac{1}{2}n_{R1} + \sum_{R1+1}^{R2-1} n_r + \frac{1}{2}n_{R2}}{\log(R2/R1)} \qquad (2.15)$$

where n_r = the total number of species with abundance R;
S = the total number of species in the sample;
$R1$ and $R2$ are the 25% and 75% quartiles;
n_{R1} = the number of individuals in the class where $R1$ falls;
n_{R2} = the number of individuals in the class where $R2$ falls.

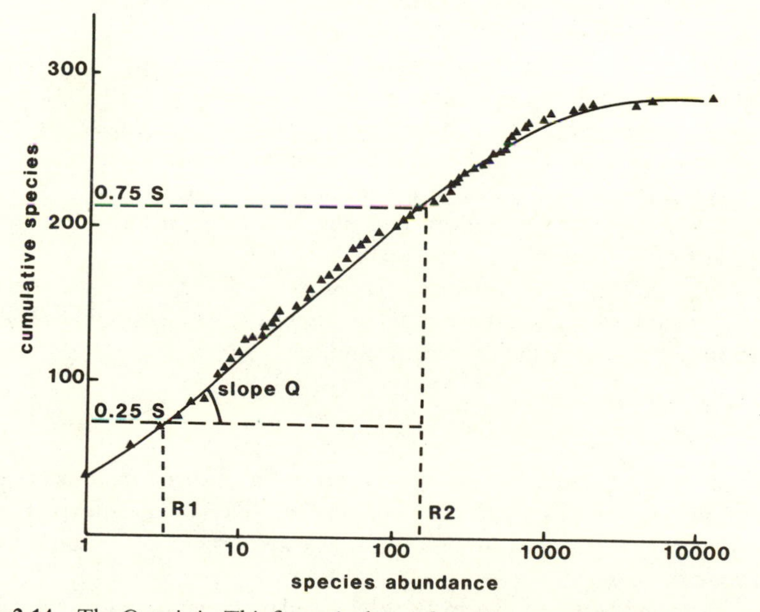

Figure 2.14 The Q statistic. This figure (redrawn from Kempton and Wedderburn, 1978) illustrates how the Q statistic is calculated. The x-axis shows species abundance on a logarithmic ($\log_{10}$) scale while the cumulative number of species is given on the y-axis. $R1$, the lower quartile, is the species abundance at the point at which the cumulative number of species reaches 25% of the total. Likewise $R2$, the upper quartile, marks the point at which 75% of the cumulative number of species is found. The Q statistic is a measure of slope Q between these quartiles.

The quartiles are chosen so that:

$$\sum_{1}^{R1-1} n_r < \frac{1}{4}S \leqslant \sum_{1}^{R1} n_r \text{ and } \sum_{1}^{R2-1} n_r < \frac{3}{4}S \leqslant \sum_{1}^{R2} n_r \qquad (2.16)$$

A worked example is shown in Example 6 (page 142).

Kempton and Wedderburn (1978) point out that Q, expressed in terms of the log series model, is analogous to α. For the log normal model $Q=0.371\ S^*/\sigma$.

Although Q may be biased in small samples, this bias is low if >50% of all species present are included in the sample (Kempton and Wedderburn, 1978).

Indices based on the proportional abundances of species

While species abundance models provide the fullest description of diversity data they are dependent on some fairly tedious model fitting and for rapid calculation require the use of a computer. In addition problems may arise if all the communities studied do not fit one model and it is desired to compare them by means of a diversity index.

Indices based on the proportional abundances of species provide an alternative approach to the measurement of diversity. Peet (1974) terms these indices heterogeneity indices because they take both evenness and species richness into account. The fact that no assumptions are made about the shape of the underlying species abundance distribution leads Southwood (1978) to refer to them as non-parametric indices. This type of diversity measure has enjoyed a great deal of popularity in recent years.

Two categories of non-parametric indices will be examined. Measures derived from information theory will be discussed first. This will be followed by an investigation of the dominance indices.

Information statistic indices

The most widely used measures of diversity are the information theory indices. These indices are based on the rationale that the diversity, or information, in a natural system can be measured in a similar way to the information contained in a code or message.

Shannon and Wiener independently derived the function which has become known as the Shannon index of diversity. It is sometimes incorrectly referred to as the Shannon–Weaver index (Krebs, 1985). The Shannon index assumes that individuals are randomly sampled from an 'indefinitely large' (that is an effectively infinite) population (Pielou, 1975). The index also assumes that all species are represented in the sample. It is calculated from the equation:

$$H' = -\Sigma p_i \ln p_i \tag{2.17}$$

The quantity p_i is the proportion of individuals found in the ith species. In a sample the true value of p_i is unknown but is estimated as n_i/N (the maximum likelihood estimator, Pielou, 1969). Use of n_i/N as an estimate of p_i produces a biased result and strictly speaking the index should be obtained from the series (Hutcheson, 1970; Bowman *et al.*, 1971):

$$H' = -\sum p_i \ln p_i - \frac{S-1}{N} + \frac{1-\Sigma p_i^{-1}}{12N^2} + \frac{\Sigma(p_i^{-1} - p_i^{-2})}{12N^3} \tag{2.18}$$

In practice however this error is rarely significant (Peet, 1974) and all terms in the series after the second are very small indeed. A more substantial source of error comes from a failure to include all species from the community in the sample (Peet, 1974). This error increases as the proportion of species represented in the sample declines. See Example 7 (page 145) for a worked example of the Shannon index and other associated calculations.

Log_2 is often used in calculating the Shannon diversity index but any log base may be adopted. It is of course essential to be consistent in the choice of log base when comparing diversity between samples or estimating evenness using equation (2.22). There is an increasing trend towards standardizing on natural logs and it is essential to use natural logs if diversity is being estimated using the series (2. 18). Pielou (1969) lists the terms used to describe the units in which the diversity is measured. These stem from information theory and depend on the type of logs used with 'binary digits' and 'bits' for $\log_2$, 'natural bel' and 'nat' for $\log_e$ and 'bel', 'decimal digit' and 'decit' for $\log_{10}$. Few ecologists now use these terms though they do crop up in the earlier literature. It seems typical of diversity measurement that one phrase will not do if half a dozen can suffice!

The value of the Shannon diversity index is usually found to fall between 1.5 and 3.5 and only rarely surpasses 4.5 (Margalef, 1972). May (1975) has shown that if the underlying distribution is log normal, 10^5 species will be needed to produce a value of $H' > 5.0$ (Figure 2.15).

Exp H' may be used as an alternative to H'. Exp H' is equivalent to the number of equally common species required to produce the value of H' given by the sample (Whittaker, 1972). The variance of H' can be calculated:

$$\text{Var } H' = \frac{\Sigma p_i(\ln p_i)^2 - (\Sigma p_i \ln p_i)^2}{N} + \frac{S-1}{2N^2} \tag{2.19}$$

and using this method Hutcheson (1970) provides a method of calculating 't' to test for significant differences between samples.

$$t = \frac{H_1' - H_2'}{(\text{Var } H_1' + \text{Var } H_2')^{1/2}} \tag{2.20}$$

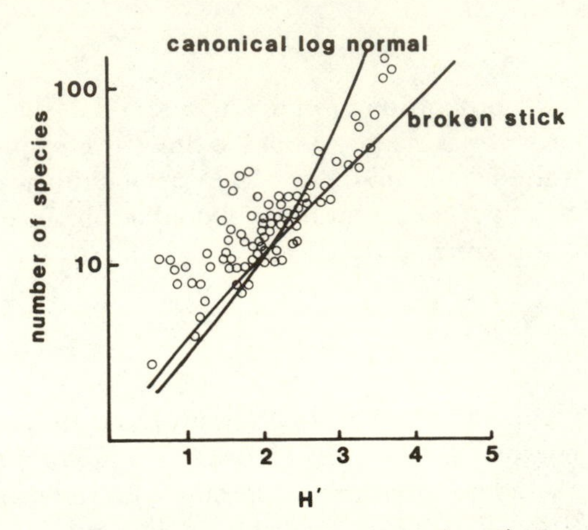

Figure 2.15 Species richness and Shannon's diversity index H'. The value of H' is related to species richness but is also influenced by the underlying species abundance distribution. In cases where this species abundance distribution is a canonical log normal, about 100 species are needed to give a value of $H' \approx 3$. For $H' > 5$ 10^5 species would be required. The dots show the relationship between H' and S for a variety of organisms (birds, copepods, corals, plankton and trees: data from Webb, 1973) and illustrate that, in the majority of cases, calculated values of H' range from 1 to 3.5. Figure redrawn from May (1975).

where, H'_1 is the diversity of sample 1 and Var H'_1 is its variance. Degrees of freedom are calculated using the equation

$$\text{df} = \frac{(\text{Var } H_1 + \text{Var } H_2)^2}{(\text{Var } H_1)^2/N_1 + (\text{Var } H_2)^2/N_2} \tag{2.21}$$

N_1 and N_2 being the total number of individuals in samples 1 and 2 respectively.

Taylor (1978) points out that if the Shannon index is calculated for a number of samples the indices themselves will be normally distributed. This property makes it possible to use parametric statistics, including the powerful analysis of variance methods (see Sokal and Rohlf, 1981), to compare sets of samples for which the diversity has been calculated (Chapter 4). This is a useful method of comparing the diversity of different habitats, especially when a number of replicates have been taken.

Although as a heterogeneity measure Shannon's index takes into account the evenness of the abundances of species (Peet, 1974) it is possible to calculate a separate additional measure of evenness. The maximum diversity (H_{max}) which could possibly occur would be found in a situation where all species were equally abundant, in other words if $H' = H_{max} = \ln S$. The ratio of observed diversity to maximum diversity can therefore be taken as a measure of evenness (E) (Pielou, 1969).

$$E = H'/H_{max} = H'/\ln S \qquad (2.22)$$

E is constrained between 0 and 1.0 with 1.0 representing a situation in which all species are equally abundant. As with H' this evenness measure assumes that all species in the community are accounted for in the sample.

Lloyd and Ghelardi (1964) have proposed a method for calculating evenness by comparing the equitability of a sample with the equitability predicted by the broken stick model. Since the broken stick model represents the most even state of affairs ever found in nature it is, they consider, a more realistic basis for estimating H_{max} than ln S. Lloyd and Ghelardi (1964) have constructed a table giving the expected number of species, derived from a broken stick distribution, for values of H'. The ratio of expected number of species against the recorded number of species is used as an index of evenness, termed J.

Lloyd *et al.* (1968) used Lloyd and Ghelardi's method to calculate the equitability of reptilian and amphibian species in the Bornean rain forest. The result they obtained was $J = 0.334$, an unexpectedly low figure and one which they found surprising in a tropical community. If however the evenness of the Bornean reptiles and amphibians is recalculated using E (where $E = H'/\log_2$) equitability doubles to 0.666. The discrepancy between the results calculated for the same data illustrates the need for caution in the use and interpretation of the deceptively simple evenness measures.

When the randomness of a sample cannot be guaranteed, as for instance during light trapping (Southwood, 1978) where different species of insect are differentially attracted to light, or if the community is completely censused with every individual accounted for, the Brillouin index (HB) is the appropriate form of the information index (Pielou, 1969, 1975). It is calculated using the formula

$$HB = \frac{\ln N! - \Sigma \ln n_i!}{N} \qquad (2.23)$$

and again rarely exceeds 4.5. Both indices give similar (and often correlated, see page 75) estimates of diversity. However when the diversity of a particular data set is estimated using both indices the Brillouin index produces a lower result (Table 2.1). This is because there is no uncertainty in the Brillouin index: it describes a known collection. The Shannon index by contrast has to estimate the diversity of the unsampled as well as the sampled portion of the community. One major difference between the indices is that the Shannon index will always give the same value providing the number of species and their proportional abundances remain constant (Table 2.1). This is not a property of the Brillouin index. Evenness (E) for the Brillouin diversity index is obtained from:

$$E = \frac{HB}{HB_{max}} \qquad (2.24)$$

Table 2.1 A comparison of the values of the Shannon and Brillouin indices.

(a) When used to estimate diversity of a single data set the Shannon index will always produce a higher value. The abundance of caddis flies collected in a light trap in Illinois. Data from Poole (1974).

Species	*Number of individuals*
Popamyia flava	235
Hydropsyche orris	218
Cheumatopsyche analis	192
Ocestis inconspicua	87
Hydropsyche betteni	20
Athripsodes transversus	11
Leptocella candida	11
Leptocella exquisita	8
Cheumatopsyche campyla	7
Polycentropus cinereus	4
Ocestus cinereus	3
Nyctiophylax vestitus	2
Cheumatopsyche aphanata	2
Neureclepsis crepuscularis	1
Triaenodes aba	1
Shannon diversity $H'=1.69$	
Brillouin diversity $HB=1.65$	

(b) The Shannon index, unlike the Brillouin index, does not vary providing the number of species and their relative proportions remain constant.

	Number of individuals	
	Sample 1	*Sample 2*
	10	5
	10	5
	10	5
	10	5
	10	5
	10	5
	10	5
	10	5
	10	5
	10	5
Shannon H'	2.30	2.30
Brillouin HB	2.13	2.01

where HB_{max} is calculated as:

$$HB_{max}=\frac{1}{N}\ln\frac{N!}{\{[N/S]!\}^{s-r}\cdot\{([N/S]+1)!\}^{r}} \tag{2.25}$$

with $[N/S]$ = the integer of N/S, and,
$r=N-S[N/S]$.

As collections, not samples, are being compared each value of HB is automatically significantly different from any other. Example 8 (page 150) provides a worked example.

Laxton (1978), investigating the mathematical properties of the index, found it theoretically the most satisfactory of the two information measures of diversity. Pielou (1969, 1975) argues strongly for its use in all circumstances where a collection (that is a non-random sample) is made or the full composition of the community known. Pielou's advice is rarely followed however as the Brillouin index is very time-consuming to calculate and can give misleading answers due to its dependence on sample size. Most ecologists using information theory measures of diversity prefer the Shannon index for its computational simplicity.

Dominance measures

The second group of heterogeneity indices are referred to as dominance measures since they are weighted towards the abundances of the commonest species rather than providing a measure of species richness. One of the best known of these is the Simpson's index. It is occasionally called the Yule index since it resembles the measure G. U. Yule devised to characterize the vocabulary used by different authors (Southwood, 1978).

Simpson's index (D) Simpson (1949) gave the probability of any two individuals drawn at random from an infinitely large community belonging to different species as:

$$D=\Sigma p_i^2 \tag{2.26}$$

where p_i = the proportion of individuals in the ith species. In order to calculate the index the form appropriate to a finite community is used:

$$D=\sum\left(\frac{n_i(n_i-1)}{N(N-1)}\right) \tag{2.27}$$

where n_i = the number of individuals in the ith species and N = the total number of individuals.

As D increases, diversity decreases and Simpson's index is therefore usually expressed as $1-D$ or $1/D$. Simpson's index is heavily weighted towards the most abundant species in the sample while being less sensitive to species

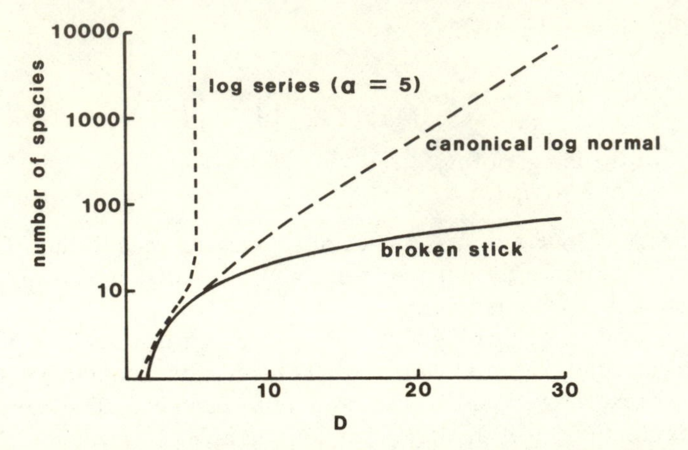

Figure 2.16 The relationship between Simpson's index, D, and species richness is strongly influenced by the underlying species abundance distribution. In situations where species abundances follow a log series distribution, Simpson's index is very insensitive to species richness. In this example where $\alpha = 5$ Simpson's index shows no increase once S exceeds 10. The other extreme occurs if species abundances are much more even and match a broken stick distribution. Here D rises dramatically with any increase in species richness above 10. With a canonical log normal distribution D displays an intermediate dependence on S.

richness (Example 9, page 152). May (1975) has shown that once the number of species exceeds 10 the underlying species abundance distribution is important in determining whether the index has a high or low value (Figure 2.16).

McIntosh's measure of diversity McIntosh (1967) proposed that a community could be envisaged as a point in an S dimensional hypervolume and that the Euclidean distance of the assemblage from the origin could be used as a measure of diversity. This distance is known as U and is calculated as

$$U = \sqrt{\Sigma n_i^2} \tag{2.28}$$

The McIntosh U index is not in itself a dominance index (see Chapter 4). However a measure of diversity (D) or dominance which is independent of N may also be calculated

$$D = \frac{N - U}{N - \sqrt{N}} \tag{2.29}$$

with a further evenness measure obtained from the formula (Pielou, 1969)

$$E = \frac{N - U}{N - N/\sqrt{S}} \tag{2.30}$$

See Example 10 (page 154).

Berger–Parker index d An intuitively simple dominance measure is the Berger–Parker index d (Berger and Parker, 1970; May, 1975). It also has the virtue of being easy to calculate. The Berger–Parker index expresses the proportional importance of the most abundant species

$$d = N_{max}/N \tag{2.31}$$

where N_{max} = the number of individuals in the most abundant species (Example 11, page 156). As with the Simpson index the reciprocal form of the Berger–Parker index is usually adopted so that an increase in the value of the index accompanies an increase in diversity and a reduction in dominance.

This index is independent of S but is influenced by sample size. May (1975) concludes that it is one of the most satisfactory diversity measures available.

Relationship between indices

Working from the observation that diversity measures can be arranged by their propensity to emphasize either species richness (weighting towards uncommon species) or dominance (weighting towards abundant species), Hill (1973) has produced an elegant method for describing the relationship between diversity indices. By defining a diversity index as 'the reciprocal mean proportional abundance' he was able to classify them according to the weighting they give to rare species. In the general case

$$N_a = (p_1^a + p_2^a + p_3^a \cdots + p_n^a)^{1/(1-a)} \tag{2.32}$$

N_a being the ath 'order' of diversity where p_n = the proportional abundance of the nth species. It follows that when $a = 0$, N_0 is the total number of species in the sample.

The orders (or numbers) of N frequently used in diversity studies are:

- $N_{-\infty}$ reciprocal of the proportional abundance of the rarest species (this is May's (1975) dimensionless ratio J)
- N_0 number of species
- N_1 exponential Shannon index
- N_2 reciprocal of Simpson's index
- N_∞ reciprocal of the proportional abundance of the commonest species (reciprocal of Berger–Parker index).

Any order of N may be employed as a diversity index but it is obviously best to use those whose properties are fairly well understood.

Hill (1973) also suggests that as the units for all the diversity numbers are the same, and that as N_a plus a constant is a good approximation to N_{a+1}, the difference between the diversity numbers might provide a plausible estimate of evenness. This is an entirely different approach to that normally adopted in measuring equitability and Peet (1974) notes that such measures can be difficult to interpret and may produce ambiguous results.

Jack-knifing an index of diversity

Jack-knifing is a technique which allows the estimate of virtually any statistic to be improved. It was originally proposed by Quenouille in 1956 with modifications by Tukey in 1958. The method was first applied to diversity statistics by Zahl (1977). Adams and McCune (1979) and Heltshe and Bitz (1979) have also investigated its effectiveness in this context.

The beauty of the method is that it makes no assumptions about the underlying distribution. Instead, a series of jack-knife estimates and pseudovalues are produced. These pseudovalues are normally distributed and their mean forms the best estimate of the statistic. Confidence limits can also be attached to the estimate.

The procedure (illustrated in Example 12, page 158) entails repeatedly recalculating the standard estimate V (for example the Shannon index) missing out each sample in turn. Each recalculation produces a jack-knife estimate VJ_i. In diversity data n jack-knife estimates will be obtained. For each sample a pseudovalue (or VP) is then calculated:

$$VP_i = (nV) - [(n-1)(VJ_i)] \quad (2.33)$$

The best estimate of V is the mean of the pseudovalues VP, and the difference between VP and V gives the 'sample influence function' which is a measure of the effect which sampling has had on the accuracy of the unjack-knifed estimate. The standard error can be obtained from

$$\text{standard error of } VP = \text{var}(VP)/S \quad (2.34)$$

with degrees of freedom equal to the number of samples minus one.

A single sample may also be jack-knifed. In this context, opinions differ as to whether $S-1$ or $n-1$ degrees of freedom should be used in the calculation of confidence intervals (Schucany and Woodward, 1977). After a Monte Carlo simulation Adams and McCune (1979) concluded that some (unspecified) function of both n and S is appropriate. They found that for 95% limits $S-1$ gave a 2–4% overcoverage while $n-1$ produced a 5–6% undercoverage. Since 't' (from t-tables) is virtually constant once degrees of freedom exceed 100 the problem will be negligible in large data sets. In smaller data sets $S-1$ will give the more conservative result and so should be favoured in most cases. In extremely small samples ($n<15$) Adams and McCune found that their attempts to set confidence limits produced erratic results. It would therefore be unwise to attach confidence intervals in similarly restricted data sets. However since sample sizes are rarely so small this limitation is unlikely to cause problems.

Investigations of the jack-knife method applied to diversity statistics have concentrated on the Simpson and Shannon indices and the conclusions drawn from these studies are most encouraging. Zahl (1977) showed that for these indices the pseudovalues are indeed normally distributed in the majority of cases. He also noted that random sampling of individuals (which is often difficult to achieve, see Chapter 3) is not required. Adams and McCune (1979) concluded that variance of the pseudovalues is 'overwhelmingly superior' to other estimates of the variance of Shannon's index (including equation 2.19). Heltshe and Bitz (1979) found that the bias of jack-knife estimates is substantially smaller than that associated with Pielou's (1969, 1975 and see Chapter 3) pooled quadrat method. In the light of these results there appears to be no reason why the jack-knife method could not be equally successfully applied to some other indices of diversity.

There is however one word of caution which must be appended to this advocation of jack-knifing. Jack-knifing a measure such as the Shannon or Simpson indices may occasionally generate a value which is patently absurd. In this instance, and in the interpretation of diversity measures generally, the results of calculations should not be followed blindly. Sophisticated mathematics are useless unless the ecologist has the skill to interpret the results in the context of the ecology of the community under investigation.

Hierarchical diversity

One final, but rarely considered, variety of diversity concerns taxonomic differences at other than the species level. Pielou (1975) points out that in intuitive terms diversity will be higher in a community in which the species are divided amongst many genera as opposed to one where most species belong to the same genus. Likewise the diversity of a particular species would be higher in a situation where there were many isolated genetically variable populations. She formalizes this concept in a version of the Shannon index which incorporates familial, generic and species diversity and shows how the same idea can be extended to the Brillouin index. Since neither diversity measure is found to be particularly easy to interpret in its species-only form (Chapter 4) it is unlikely that their extension upwards to generic and familial diversity or downwards to population diversity will prove informative in the vast majority of cases. For simplicity of calculation and interpretation it would seem preferable to use a genus or family richness measure (calculated on the same lines as species richness) in those studies in which a perspective on hierarchical diversity is desired. Alternatively, it may be more satisfactory to abandon taxonomy altogether and record instead the diversity of growth forms (Harper, 1977).

Further reading

Other reviews of diversity measures and models are provided by Peet (1974), May (1975)*, Pielou (1975)*, Engen (1978)*, Southwood (1978), Grassle *et al.* (1979)*, Frontier (1985) and Gray (1988). Readers seeking a fuller mathematical treatment should consult the starred reviews and follow up the original papers which can be accessed via the references scattered through the text.

Calculations

It is possible to do all the mathematical calculations described in this chapter using a pocket calculator with scientific functions. The examples illustrate the procedures involved for the majority of indices and measures. While it is valuable to do each mathematical procedure at least once by hand, computers greatly speed the operation. None of the calculations are difficult to program on a micro-computer and the cook-book format of the Examples should assist in this process. (It was decided to include 'recipes' for the methods rather than actual programs since ecologists use a diversity of programming languages!) The truncated log normal is the only potential source of trouble since the calculations involved in obtaining the auxiliary estimation function θ are complex. The simple solution is to skip this step and take the value from Cohen's (1961) table, reproduced in Appendix 2. Readers with access to mainframe computers may find that some programs to fit the more common models already exist.

There is always a great temptation to try out more and more indices and models on a set of diversity data. In most cases it is most economical and informative to restrict the analysis to just one or a few of the more commonly adopted measures. Chapter 4 makes specific recommendations.

Summary

This chapter has reviewed the many diversity measures and models that ecologists use. These can be divided into three major groups: the species richness measures, the species abundance models (some of which have associated diversity indices) and the indices which are based on the proportional abundance of species. Species richness indices, for example the species count and the Margalef index, are intuitively simple but sensitive to sample size.

By looking at the full species abundance distribution it is possible to get a better picture of the relationship between species richness and evenness, that is the relative abundances of the species present. A number of models have been proposed to account for different species abundance patterns but often the biological assumptions on which these are based are discredited or unproven. It is more useful to use models as statistical fits to empirical data. In this way it is possible to trace a sequence from the geometric series, which reflects a situation in which one or a few species are dominant, the rest rare, through the log series and log normal to the broken stick which represents the greatest degree of evenness, that is the greatest equality in species abundances, found in nature.

Indices based on the proportional abundances of species offer a half-way house. Some of these, for instance the Berger–Parker index which measures dominance, are simple to use and informative, while others, for example the popular Shannon index, are more difficult to interpret. Hill (1973) shows how diversity indices are mathematically related and can be arranged in a sequence according to whether they measure richness or dominance.

The procedure of jack-knifing, which is a method of improving the estimate of a diversity index, is described briefly.

3 Sampling

It is rarely feasible, or desirable, to census every individual in a community. Such a strategy would be prohibitively time-consuming and expensive; it would also damage or possibly even destroy the community in question. Ecologists therefore rely on sampling to provide an accurate picture of community composition. A great deal of effort over past decades has been devoted to making sampling techniques as efficient as possible. Southwood (1978) for instance describes the various approaches to sampling insect populations while Kershaw and Looney (1985) and Moore and Chapman (1985) discuss the methods available for the sampling of plant communities. Diversity studies raise a number of special problems where sampling is concerned. For example can individuals be sampled randomly? What size should samples be? What happens if individuals are not easily recognizable? How should a community be defined? This chapter discusses these problems and provides some suggestions for solving them.

Random sampling?

Most sampling methods can be adapted to provide a random coverage of the study area. For instance pitfall traps can be sited using random number tables, quadrats for recording ground vegetation can be placed on the basis of a random walk and so on. Elliot (1977), Lewis and Taylor (1967) and Southwood (1978) are but three of the many texts which give advice on random sampling. But random coverage of an area is not in itself random sampling of individuals. A whole host of reasons including predator avoidance, competition, habitat requirements and modular growth form (see Krebs and Davies, 1981, 1984; Hassell and May, 1985; Harper, 1977, 1981) lead organisms to aggregate (see Figure 3.1). When this occurs it is 'probably impossible' (Pielou, 1975) to ensure that individuals will be randomly sampled even when the sampling device is itself randomly positioned. This non-randomness is important because diversity indices assume that the probability of two successively sampled individuals belonging to the same species is

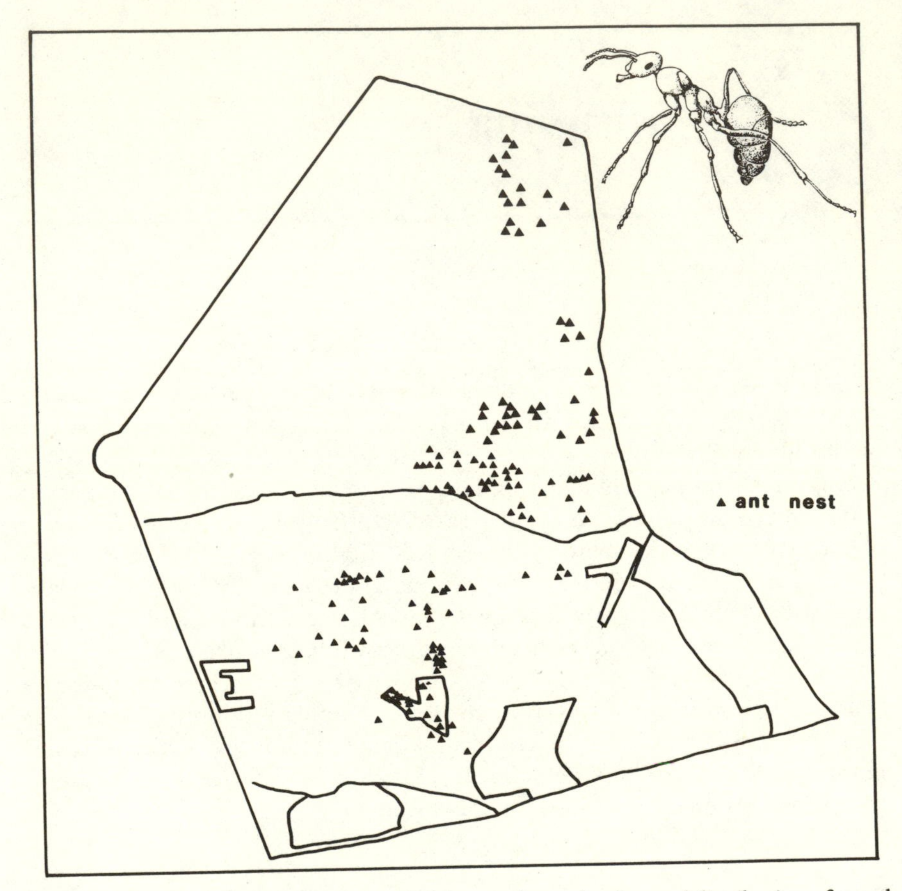

Figure 3.1 Aggregation of organisms. This map shows the clumped distribution of wood ant nests in Bedford Purlieus. Redrawn from Peterken (1981).

dependent only on the relative abundances of species within the community. De Caprariis and Lindemann (1978) show that aggregation affects even species richness estimates.

Jack-knifing the estimate of a diversity index (see Chapter 2) is one simple solution to this problem. This technique is robust against bias caused by clumping (Zahl, 1977). Precision will be further increased by ensuring that the quadrats or other sampling units are placed at random and that a reasonably large sample size has been taken (see under sample size).

Pielou's (1966, 1969, 1975) pooled quadrat method provides an alternative method of circumventing the problem. It works as follows. A series of randomly placed samples are taken, pooled in random sequence and the cumulative diversity calculated using the Brillouin index (Figure 3.2). The

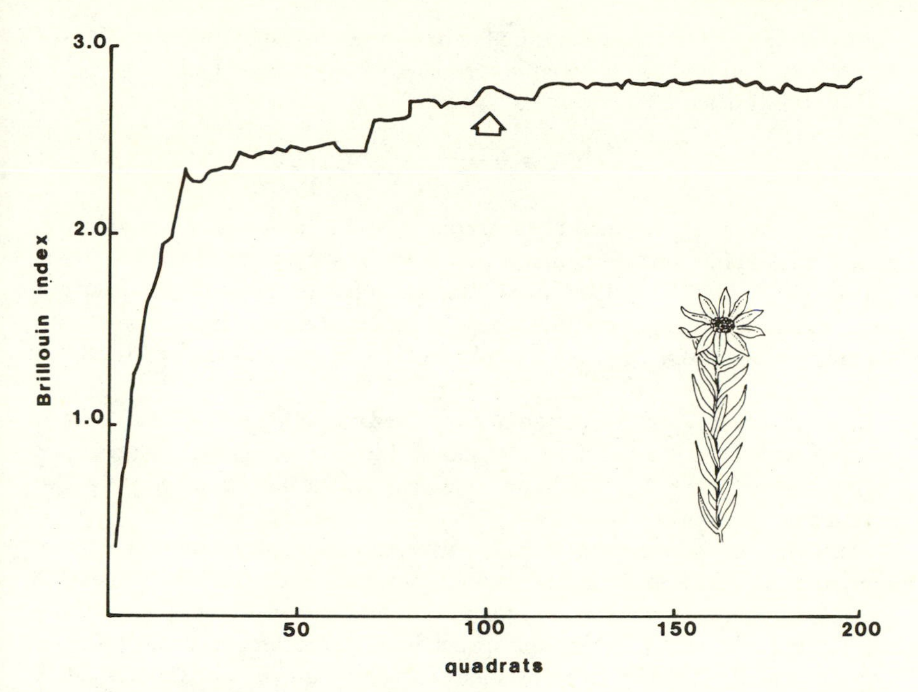

Figure 3.2 A Brillouin cumulative diversity curve to show the diversity of ground vegetation in Banagher conifer plantation, Northern Ireland (Figure 4.2). Data were collected using 200 point quadrats. These quadrats were then pooled in a random sequence and the Brillouin index continuously recalculated. The resulting curve flattens at around the 100 quadrat point (shown by the arrow). This flattened portion of the curve would be used to calculate diversity. See text and equation (3.1) for details.

Brillouin index is chosen since Pielou considers it to be more appropriate than Shannon's index which assumes a truly random sample. The Brillouin cumulative diversity HB_k is plotted against the number of quadrats, k. The point at which the resultant curve flattens off is referred to as t and the flattened portion of the curve is used to estimate population diversity, HB_{pop}. To do this, values of h_k from $k=t+1$ to $k=z$ (where $z=$ the total number of quadrats or samples) are calculated from the formula:

$$h_k = \frac{M_k HB_k - M_{k-1} HB_{k-1}}{M_k - M_{k-1}} \qquad (3.1)$$

where $HB_k=$ the diversity of the kth (cumulative quadrat) calculated using the Brillouin index (see Chapter 2) and $M_k=$number of individuals or other biomass measure in the kth cumulative quadrat.

Although there is an element of subjectivity in deciding the point at which the curve flattens off, t can be assumed to have been chosen correctly if values of h_k between t and z are not serially correlated (Poole, 1974; Pielou, 1975).

H_{pop} is estimated by

$$H_{pop} = \frac{1}{z-(t+1)} \sum_{k=t}^{z} h_k \qquad (3.2)$$

The variance of H_{pop} is equal to the variance of the values of h_k divided by n, where n is the number of estimates of h_k and the variance is calculated in the usual way (Poole, 1974). Confidence limits can be attached to the estimate of H_{pop}. Example 13 (page 160) illustrates the procedure. Note however that this variance is less satisfactory than that obtained from the jack-knife method (see Chapter 2).

For any data set each calculation of H_{pop} based on a different random order of samples will produce a different estimate. Lloyd *et al.* (1968) suggest that 'several' estimates of H_{pop} should be calculated and the median taken as the best estimate of population diversity.

This method of estimating diversity requires a considerable amount of computation. Brillouin's index, which involves factorials, is tedious to calculate and the sorting and sequential accumulation of samples is time-consuming, especially when the procedure is repeated several times as recommended. For these reasons Pielou's pooled quadrat method is rarely adopted. In practice estimates of H_{pop} are highly correlated with estimates of diversity made using the Shannon and other indices which have been calculated without the strict interpretation of random sampling. For instance, Magurran (1981) estimated the diversity of vegetation in ten woodlands (see Figure 6.2) and found highly significant correlations ($P<0.01$) between the values of H_{pop} and the standard Brillouin index ($r_s=0.93$), the Shannon index ($r_s=0.93$), species richness ($r_s=0.92$) and the Margalef index ($r_s=0.96$).

The greater computational efficiency and accuracy of the jack-knife method means that it is the preferable technique, especially when the species in the study are known to have a clumped distribution. Nevertheless the pooled quadrat method can be used with other diversity indices (Heltshe and Bitz, 1979) and it is also a useful way of deciding the appropriate sample size (see below).

Other factors can lead to non-randomness in a sample. Southwood (1978) provides a good review of possible sources of bias. Different groups of insects differ in their susceptibility to light traps for instance and the positioning of the trap itself can critically affect its attractiveness. Work by Taylor and French (1974) has shown that if the diversity of moths in different sites is being compared by light trapping it is essential that the light traps are of the same design, that they are sited so that they are equally visible and that they are placed at a constant height above the ground. Weather conditions must be taken into consideration too. Cold, wet, windy and moonlit nights tend to produce low catches (Holloway, 1977) and unless traps are run simultaneously differences between sites can be obscured. Seasonality is obviously another important problem.

Different sources of bias will be associated with different types of trap or sampling device and different groups of animals and plants. It follows that in any survey the ecologist should be aware of this bias and should understand as fully as possible the behaviour and ecology of the organisms being sampled. This may be an elementary point but it is fundamental to the successful study of ecological diversity. A recent handbook by Chalmers and Parker (1986) provides sound advice on a variety of ecological fieldwork techniques.

Sample size

One problem associated with diversity measurement is knowing what sample size to adopt. In practice most people take the pragmatic approach and sample until time or money runs out or until they intuitively feel that they have adequately described the diversity. If species richness alone is being measured the problem is rather simpler and as soon as the boundaries of the community have been defined (see below) it is necessary only to record species presence. Sampling intensity however affects even species richness. In the Rothamsted insect survey (Taylor, 1986) light trapping for moths over successive years has added more and more new species (usually vagrants) to the species total, and Connor and Simberloff (1978) found that the number of botanical collecting excursions to the Galapagos Islands was a better predictor of species richness than area or isolation. Kirby *et al.* (1986) showed how the number of vascular plants recorded in a broadleaved woodland increased with survey effort (Figure 3.3).

Pielou's pooled quadrat method can be usefully adapted to provide a guide to sample size. As before quadrats (or other sampling units) are pooled in random order and diversity continuously recalculated on the basis of all the data currently in the pool. The point at which the curve flattens indicates the minimum viable sample size. Any diversity index or indices can be used. Hill's (1973 and see Chapter 2) family of diversity measures are, as we have already seen, a valuable way of focusing on different aspects of the species abundance distribution (Kempton, 1979) making it possible to emphasize either the degree of dominance or the contribution of rare species. Since Hill's diversity numbers are expressed in the same units two or more diversity curves can be plotted simultaneously to this end. The diversity curve constructed using N_0 (that is S, the number of species) is equivalent to the conventional species area curve (Hopkins, 1957).

The choice of index will govern the computational complexity of the diversity curve. For indices such as Shannon or Simpson where hand

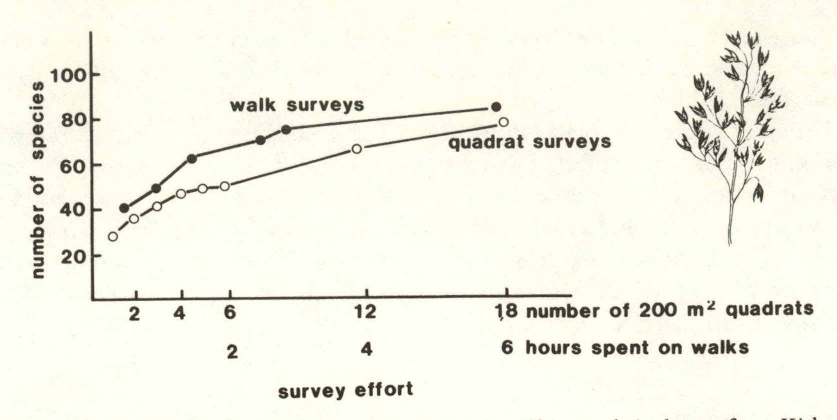

Figure 3.3 Diversity is related to sampling intensity. This graph (redrawn from Kirby *et al.*, 1986) shows the relationship between the number of vascular plant species recorded and sampling effort, in walk surveys and quadrat surveys carried out in a broadleaved wood in April.

calculation is fairly laborious a computer program is desirable. However the simple indices such as N_0 and the Margalef index (which both measure species richness) and the Berger–Parker index (which measures dominance) permit the rapid construction of a diversity curve, in the field if need be.

Two sets of diversity curves, based on Hill's measures and describing ground flora, are illustrated in Figure 3.4. The data were collected by quadrat survey (Magurran, 1981) in two contrasting woodlands: an oakwood which is a remnant of primeval forest and also a nature reserve, and a conifer plantation (see Chapter 4). The N_0 or S curve rises steeply in both the oakwood and the plantation. This confirms the earlier observation (Chapter 2) of the dependence of S on sample size. The diversity curves produced by the other three indices level off at about 50 quadrats in both sites indicating that this is the minimum sample size on which a diversity estimate should be based.

Table 3.1 lists the estimates of diversity made using a variety of indices for two independent sets of 100 random quadrats from the two woods. The first point to note from this table is that the two separate estimates of diversity for each site yield very similar results for the same number of quadrats.

The second point is that there is a difference in diversity as estimated at 50 and then at 100 quadrats. This difference varies between indices and also between sites. The Brillouin index and the exponential Shannon index are stable in the plantation but increase in the oakwood. The indices sensitive to dominance (Berger–Parker and Simpson) decrease in the plantation but do the reverse in the oakwood. Like the information statistics the Margalef index rises in the oakwood but remains stable in the plantation. For these reasons it is essential that the same sample size should be used in all sites under investigation. This conclusion is supported by Figure 3.5 which shows the confusion that can

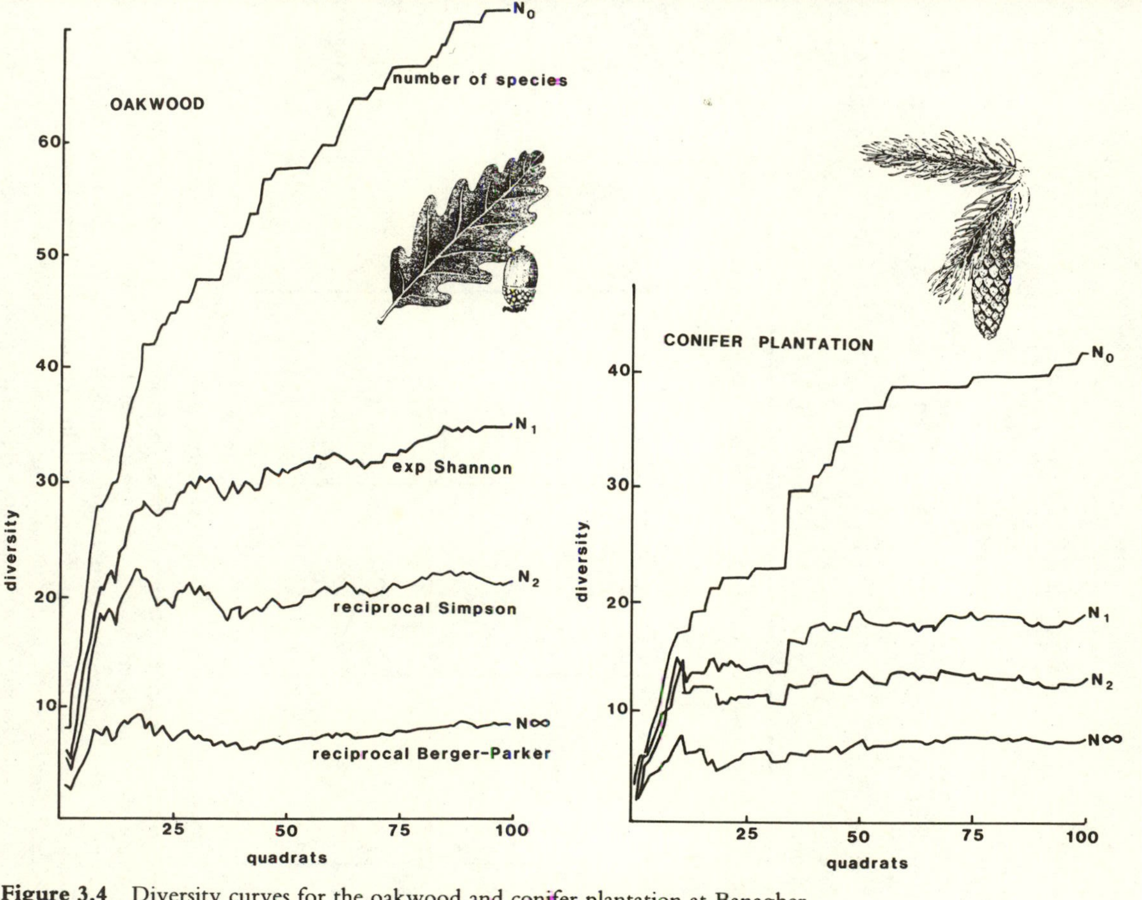

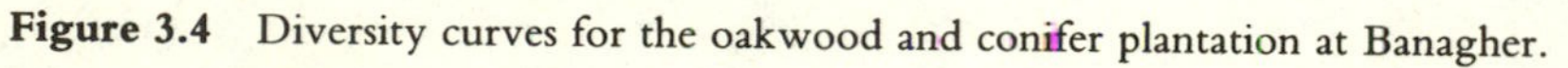

Figure 3.4 Diversity curves for the oakwood and conifer plantation at Banagher.

Table 3.1 Estimates of diversity of ground flora based on two sets of 50 and 100 random quadrats from an oakwood and a coniferous plantation in Northern Ireland. The results are derived from the reciprocal forms of the Berger–Parker (N_∞) and Simpson ($1/D$) indices and the exponential form of the Shannon index (exp H). The number of 'individuals', the abundance measure, was estimated by counting the total touches of a point quadrat to the vegetation.

	Coniferous plantation				*Oakwood*			
Data set	*Set 1*		*Set 2*		*Set 1*		*Set 2*	
Quadrats	50	100	50	100	50	100	50	100
Species richness	37	42	32	48	57	75	58	73
Individuals	404	772	396	771	858	1816	931	1815
Margalef	6.0	6.2	5.2	5.2	8.3	9.9	8.3	9.6
Berger–Parker	7.2	7.5	9.0	7.6	7.4	10.1	6.8	8.5
Simpson	13.6	13.1	14.0	13.3	17.9	20.3	19.4	21.4
Shannon	19.2	19.1	18.1	18.8	29.2	33.4	31.1	35.1

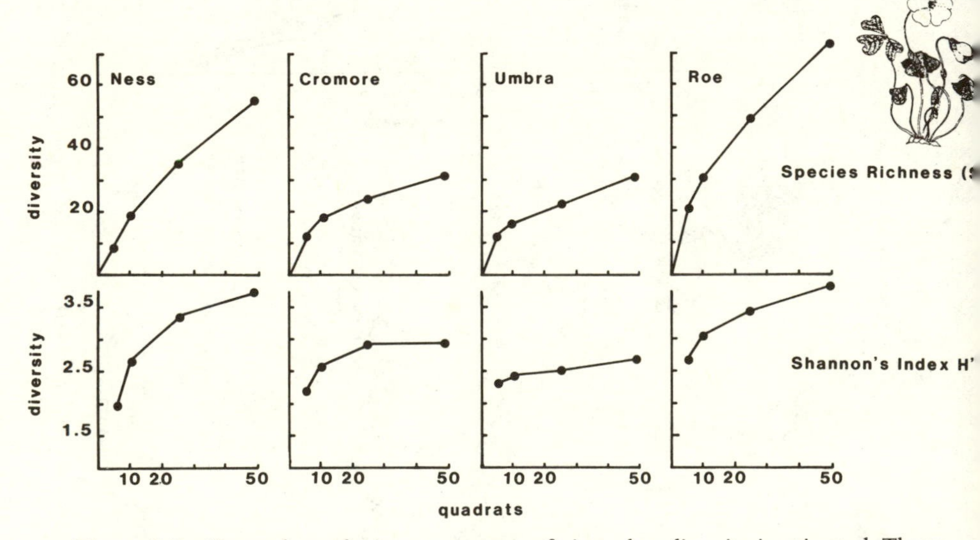

Figure 3.5 Unequal sample sizes can cause confusion when diversity is estimated. These graphs show the diversity of ground vegetation in four woodlands (see Figure 6.6) estimated on the basis of five, ten, 25, and 50 quadrats. Two diversity measures, species richness, S, and Shannon's index, H', are used. In each case a misleading result would have been obtained if a large sample (50 quadrats) had been taken in the species-poor woods (Cromore and Umbra) but only a small one (five or ten quadrats) in the species-rich woods (Ness and Roe).

result when diversity is compared in samples of different sizes. Similarly Minshall *et al.* (1985) found that a survey of benthic invertebrates based on the examination of 10 rocks in each site along the Salmon River, Idaho, USA, yielded only 60–90% of the species found when 35 or 40 rocks were examined. In order to establish sample size it is therefore advisable to construct a diversity curve for what is considered likely to be the most diverse site and plan the sampling regime accordingly.

In situations where sample sizes are unequal rarefaction and allied techniques (see Chapter 2) can be used to reduce all samples to a standard size.

The size of basic sampling unit, for example the quadrat, should be chosen according to the nature of the organisms being investigated. Guidelines for doing this are to be found in standard methods texts [see for example Chalmers and Parker (1986), Southwood (1978) and Russell and Fielding (1981)]. As a general rule a large number of small quadrats is preferable to a small number of large quadrats.

The case of the indiscrete individual

Diversity indices and species abundance models were largely developed using data from groups of animals such as moths and birds where individuals are readily identifiable. In many situations however it is difficult to decide where one individual ends and the next one begins. Plant communities for example may contain many clonal species in which a single individual can cover a considerable area simply by repeating the modular unit (Harper, 1977). Harberd (1967) showed that one genetic individual of the grass *Holcus mollis* extended over a kilometre despite being fragmented into a number of phenotypic units. Although it is often possible literally to unearth the extent of a clone by excavating its root system it takes only a moment's reflection to see that such drastic and destructive action would not provide a meaningful measure of abundance to plug into an index or model. Resource apportioning theory assumes that abundance is in some way proportional to niche size (Chapter 2). Harper (1977) notes that the weights of individual plants within a species can vary 50 000-fold. This observation clearly shows that the number of individuals has no correlation with the subdivision of one niche axis, horizontal space, between species. The choice of the correct abundance measure is also relevant to other communities where there are many clonal organisms, for example littoral zones and coral reefs.

A variety of other measures of abundance can be substituted for N (number of individuals) in diversity measurement. The number of modular units per species in a plant community is one alternative (Harper, 1977). Modular units, which are relatively constant in size within a species, include the shoot of a tree, the tiller of a grass and the leaf and bud of an annual. Harper sees the number of modular units of primary use in studies of population dynamics which, by definition, are concerned with only one species. However if the species for which diversity is being measured all have a similar growth form there is no reason why modular units should not be counted in order to measure abundance.

A more universally applicable measure of abundance is biomass. This has been used successfully in many studies including those of Pielou (1966) and Kempton (1979). Biomass can be time-consuming to measure. In plant communities for instance it involves harvesting the vegetation and sorting it into species lots which are then individually dried and weighed. Despite this drawback biomass has many advantages. It is a more direct measure of resource use than number of individuals, even where individuals are easily distinguished (Harvey and Godfray, 1987). It is a continuous measure and hence more appropriate for use in conjunction with the log normal model. It is an easily understood measure and one that is readily transportable across different groups of organisms. Finally it provides a more meaningful comparison between the diversities of different taxonomic levels of organisms. While the density of a population of soil bacteria and deer in a metre square varies by over 25 orders of magnitude (respectively 10^{21} to 10^{-5} per m^2) the range of biomass of the same organisms covers only four orders of magnitude (0.001 to 1.1 g m^{-2}) (Odum, 1968). Interestingly, the variation between the microbes and mammals decreases further when an even more fundamental unit of resource use, energy flow, is considered (May, 1981). The difficulties of taking random samples of individuals has been alluded to earlier. One of the major disadvantages of using biomass as a measure of abundance is that it is well nigh impossible to sample randomly.

The area that plants or other sessile organisms cover can also be used to replace number of individuals as the abundance measure. The coverage of individual species is often expressed as a percentage of the total area surveyed. See for example De Caprariis and Lindemann (1978) who looked at the diversity of coelenterates in a coral reef off Florida, Whittaker (1965) who investigated the diversity of plant species in the Sonoran desert and Thomas and Shattock (1986) who studied the filamentous fungal associations of *Lolium perenne*. Cover can be estimated directly in the field or measured more accurately using photographs which are subsequently digitized. Problems arise when organisms overlap one another or when there is a combination of elongated species, such as grass, and prostrate species, such as bryophytes.

Although easier to use, cover scales such as those of Domin, Braun-Blanquet (Kershaw and Looney, 1985) and Daubenmire (Mueller-Dombois and Ellenberg, 1974) do not provide an adequate substitute for abundance. These scales give the greatest discrimination at maximum and minimum cover. They are not linearly correlated with abundance and as such would produce a biased result if used in conjunction with diversity models and indices.

Point quadrats have also been developed by plant ecologists to measure cover (Chalmers and Parker, 1986; Kershaw and Looney, 1985). A point quadrat consists of a frame of pins which is adjusted to be at vegetation height. The pins are then dropped one at a time and the species touched by each pin recorded. The total number of 'hits' for each species is equivalent to its abundance. Magurran (1981) found this method useful in a study of the diversity of woodland vegetation while Southwood *et al.* (1979) employed it to measure both the taxonomic and structural diversity of a secondary succession (see Chapter 5).

One other common technique of estimating abundance is frequency or incidence. The number of sampling units that a species occurs in is added to obtain its total abundance. Although there are occasions where such an approach can give a valid measure of abundance it will often lead to an underestimate of the abundance of the commonest species and should be used with discretion. For instance a species which was very widespread and covered virtually the whole area of every quadrat would be counted as being equally abundant with the species which had but a single individual in each quadrat. To circumvent this problem it would be necessary to have a large number of very small quadrats.

Hengeveld (1979) includes these alternatives in a list of 14 widely differing definitions of abundance and adds some caveats about the interpretation of abundance data. For the purposes of estimating species diversity it is obviously important to be consistent in the abundance measure used and not to mix for example biomass and cover within the one calculation.

Defining a community

So far no attempt has been made to clarify the meaning of the word community. Krebs (1985) defines a community as 'a group of populations of plants and animals in a given place' while Begon *et al.* (1986) describe it as 'an assemblage of species populations which occur together in space and time'. Southwood (1988), in a review entitled 'The concept and nature of the community', sees a community as an organized body of individuals in a specified location. In all three definitions, which are representative of the ecological literature as a whole, the idea of community is partitioned into two components. First a community is made up of a group of interacting organisms. This group may be as limited as a single guild or may embrace everything from bacteria to buffalos. Second, the community exists within defined spatial boundaries. Thus we can refer to a community of insects on a bracket fungus, a community of plants in a field or a community of plants and animals in a tropical rain forest. Lambshead *et al.* (1983) substitute the word assemblage for community. They define an assemblage of species as the result of adequate sampling of all organisms of a specific category in a defined place.

The need to define and delimit the community will arise in any investigation of ecological diversity. Whittaker's (1972, 1977) notion of *inventory* diversity helps structure this decision. Whittaker (1977) distinguishes four levels of inventory diversity. On the smallest scale is *point* diversity, the diversity of a micro-habitat or sample taken from within a homogeneous habitat. The diversity of this homogeneous habitat, the second of Whittaker's categories, is termed *alpha* diversity, and is directly equivalent to MacArthur's (1965) idea of within-habitat diversity. The next scale of inventory diversity is *gamma* diversity, the diversity of a larger unit such as an island or landscape. As gamma diversity is defined to be the overall diversity of a group of areas of alpha diversity so *epsilon* or regional diversity, the fourth category, is the total diversity of a group of areas of gamma diversity. Whittaker envisages epsilon diversity applying to large biogeographic areas.

Although Whittaker matched his categories to fairly precise scales (habitat, landscape, biogeographic area) the idea can be easily adapted. It could for example be useful to define a single plant as a unit of alpha diversity and to record the variety and abundance of insect species found on it. Linking in with this definition might be a leaf as an area of point diversity, a group of plants occurring together as an area of gamma diversity and the forest within which the plants are located as an area of epsilon diversity. Lawton (1976, 1978, 1984) has for instance looked at the variety of insects feeding on bracken (*Pteridium aquilinum*) at the level of frond, patch, country and continent while Southwood and Kennedy (1983) have worked on the theme of trees as islands. Begon *et al.* (1986) note that 'a community can be defined at any size, scale or level within a hierarchy of habitats' and give examples of three scales: the flora and fauna in a deer's gut, the beech/maple woodland within which the deer is found and the temperate forest biome of North America. Each of these can be legitimately treated as a community. Inventory diversity can be measured by any of the methods outlined in Chapter 2. The associated idea of *differentiation* diversity, which is the difference in diversity between areas of point diversity, alpha diversity or gamma diversity is examined in Chapter 5.

Hughes (1986) notes that an ecologist's view of what constitutes a community can depend on which species abundance model is preferred. Advocates of log series models may thus consider communities to be smaller, and less self-contained, entities than those who favour the log normal. This is because higher levels of extinction and immigration, and consequently a greater proportion of rare species, will arise when 'communities' consist of relatively small numbers of species. The way in which an increase in sample size can change the pattern of species abundance from log series to log normal has already been explored (see Chapter 2).

It is unlikely that any decision about the physical boundaries of the study area will be made independently from the choice of the group of organisms to be studied. The problems in carrying out a complete census of a habitat are enormous and in most cases the degree of taxonomic expertise required limits investigations to one or two groups at most. Some of the most interesting studies contrast the diversity of different organisms. For instance Southwood *et al.* (1979) concluded that insect diversity was related to plant taxonomic diversity in the early stages of a fallow field to birch woodland succession. In the later stages of the succession however plant structural diversity was more important (see Chapter 5).

It should be stressed that since the diversities of different groups of organisms within a habitat are not necessarily correlated, for example bird diversity and floristic diversity in a conifer plantation (Moss, 1978, 1979), extrapolations from one group to others should be made with great care.

Diversity measures are most informative and easiest to interpret when they are applied to fairly limited, and well defined, taxonomic groups. Thus if the diversity of a small woodland was under investigation it would be most profitable to assess the diversity of birds, butterflies, beetles and bryophytes separately.

Summary

The precise aims of each study will largely determine the extent of the study area and the taxon or taxa to be studied. If a number of communities are being compared it is vital to be consistent in the choice of sample size. It is important too that the sample size is sufficiently large to represent diversity adequately. The pooled quadrat method is one way of doing this. Since it is often difficult to ensure that samples are taken randomly the jack-knife method should be used where possible to improve the estimate of diversity. The number of individuals is an unsatisfactory measure of abundance for many organisms and a variety of alternative measures are discussed.

4
Choosing and interpreting diversity measures

Given the large number of indices and models it is often difficult to decide which is the best method of measuring diversity. One good way to get a 'feel' for diversity measures is to test their performance on a range of data sets. There are two approaches to this. First, by looking at contrived data it is possible to observe how the different measures react to changes in the two major components of diversity, species richness and evenness. However, in the real world it is rare for richness and evenness to vary independently in the way they so often do in artificial data sets. The second, and more realistic, approach therefore is to test the response of diversity measures to species abundances from genuine ecological communities. This chapter begins by comparing the behaviour of a range of diversity measures and models when used to estimate the diversity of two data sets, one contrived and one real. The difficulties of deciding the appropriateness of one species abundance distribution over another have already been mentioned (see Chapter 2) and quickly become apparent when models are fitted to data. Often the problems arise when a goodness of fit test fails to discriminate between different distributions. The value of goodness of fit tests in conjunction with, or instead of, graphical methods is considered in the context of the analysis of data sets.

A rather more scientific method of selecting a diversity index is on the basis of whether it fulfils certain functions or criteria. In the second part of the chapter diversity measures are assessed in relation to four criteria: ability to discriminate between sites, dependence on sample size, what component of diversity is being measured, and whether the index is widely used and understood.

The chapter concludes with a list of guidelines for choosing and using diversity measures.

Richness, evenness and the killer quail

An ecologist investigates the bird diversity of three little known woodlands in a remote European country. In each case the birds visible or audible from random positions along transects are counted until the total number of individuals recorded reaches 500. Rank abundance plots are constructed and diversity estimated using nine of the more popular indices. The fit, or otherwise, of the log series, log normal and broken stick models is assessed. All methods are described fully in Chapter 2.

Inspection of the data (Table 4.1) shows immediately that species richness

Table 4.1 Bird species abundance in remote European woodlands. For more details see text.

	Hidden Glen	*Wild Wood*	*Lonely Pines*
Spotted ratcatcher	1	2	0
Killer quail	3	16	354
Riff raff	2	3	7
Slyneck	1	2	4
Oat crake	4	10	29
Cold start	5	13	4
Big dipper	1	30	3
Shylark	1	14	12
Startling	18	22	18
Deadwing	1	1	2
Crook	2	4	1
Nightcap	63	5	1
Golden lover	2	19	1
Baby bunting	1	18	1
Mute swain	1	14	2
Chinese kite	1	15	0
Brownie owl	16	1	3
Hen hurrier	15	27	1
Grrrr falcon	60	36	0
Gosh hawk	1	3	2
Cough	1	47	0
Flapwing	8	38	18
Not	16	4	0
Bar-tailed nitwit	127	6	0
Snoop	9	7	0
Funny tern	18	8	1
Cut throat	3	16	0
Throttled dove	4	32	0
Ribbon	3	19	1
Backchat	11	6	1
Missile thrush	6	7	1
Cold tit	7	8	11
Twit	8	16	9
Yellow spanner	63	27	10
Born howl	17	4	3

(S) is the same in two of the three woods. Those measures which are a combination of S and N (total number of individuals), for instance the Margalef index and the log series index α, also give these woods equal diversity (Table 4.2). The rank abundance plot (Figure 4.1) however shows that Wild Wood has fewer abundant and rare species than Hidden Glen. This observation is borne out by the indices which incorporate information on the proportional abundances of species, the Shannon index, the Simpson index, the Berger–Parker index and the log normal index λ. Evenness is greater in Wild Wood and hence the bird fauna here is more diverse than that of Hidden Glen (Table 4.2). The lower dominance of Wild Wood is reflected by the finding that it is the only site adequately described by the broken stick model. Likewise the fact that the log series is appropriate to the other two woods emphasizes their lower evenness – even when, as in the case of Wild Wood and Hidden Glen, the numbers of species and individuals are identical. (The observation that the truncated log normal fits all sites will be followed up below.)

In the third woodland, Lonely Pines, species richness is low ($S=26$), and due to the abundance of the killer quail, evenness is also low. As a consequence all indices show that it is clearly less diverse than the other two sites.

What can we conclude from this exercise? In the first instance species richness, while giving a valuable insight into the bird diversity, can mask shifts in dominance/evenness. It would therefore appear important to couple an

Table 4.2 (A) The diversity of the three woods in Table 4.1 calculated using a variety of diversity statistics, and (B) the pattern of species abundances in the three woods. The 'fit' of three species abundance distributions, log series, truncated log normal and broken stick, is tested using the methods described in Chapter 2. The critical value of P in the χ^2 goodness of fit test is $P>0.05$.

	Hidden Glen	*Wild Wood*	*Lonely Pines*
(A) Diversity			
Species richness (S)	35	35	26
Individuals (N)	500	500	500
Margalef	5.47	5.47	4.02
Berger–Parker (N_∞)	3.49	10.64	1.41
Simpson ($1/D$)	8.50	21.86	1.97
Shannon	2.61	3.23	1.38
Shannon evenness	0.74	0.91	0.42
Log series index (α)	8.57	8.57	5.82
Log normal index (λ)	53.41	78.14	43.67
(B) Fit of models			
Log series	Yes	No	Yes
Log normal	Yes	Yes	Yes
Broken stick	No	Yes	No

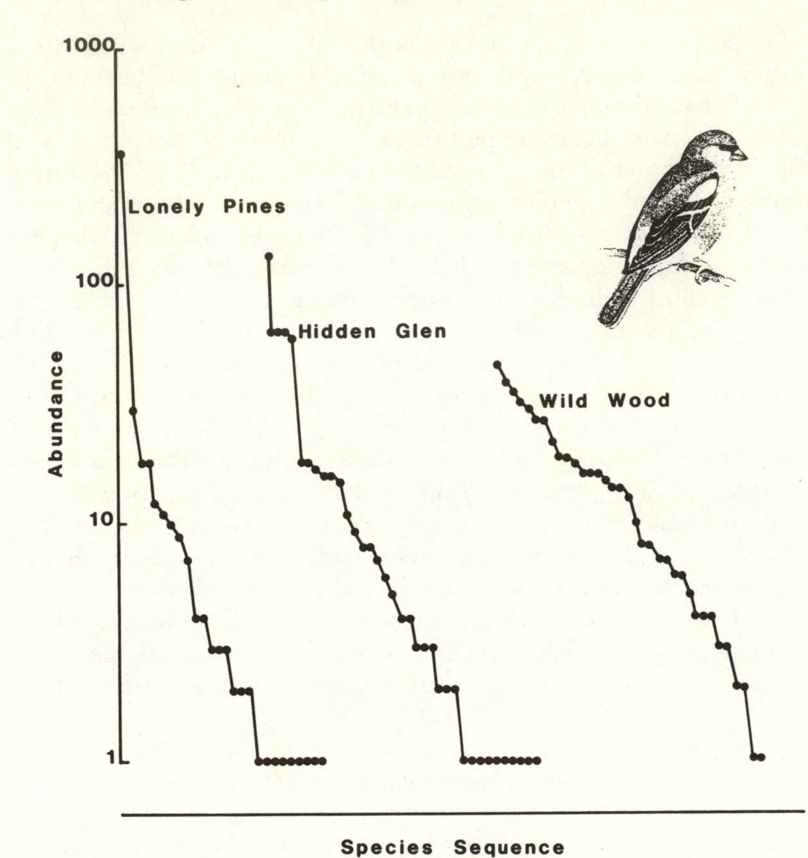

Figure 4.1 Rank abundance plots of the data in Table 4.1.

estimate of species richness with a measure of either dominance or evenness wherever possible. The Berger–Parker index seems ideal for this function. Together these measures (Berger–Parker and S) are simpler to calculate and more informative than either the Shannon or Simpson measures. Like the Margalef index the log series α fails to discriminate situations where S and N are identical but evenness varies. Although this phenomenon is unlikely to occur in genuine data sets it can easily be detected by judicious use of a Berger–Parker style index. The species abundance distributions confirm the patterns of dominance and evenness revealed by the various indices. Finally, it may be prudent to learn more about the ecology of particularly common species. In this instance the killer quail is an obvious contender for further investigation.

An oakwood and a conifer plantation in Ireland

During the last two decades the environmental lobby in Britain and Ireland has expressed considerable concern over the expansion of conifer plantations. Plantations are the least attractive form of woodland for conservation (Peterken, 1981). Yapp (1979) has said with reference to the claim that planting is unfavourable to wildlife that 'this must always be a subjective judgement, but measurement of an index of diversity for different groups of animals and plants can help to provide the necessary facts on which such a judgement can be based'. So how well do diversity measures perform in this context and which is the best index to use? In order to test Yapp's proposition we compare the diversity of two groups of organisms, ground vegetation and macro-lepidoptera, in two very different types of woodland in Ireland.

Situated at Banagher in the Sperrin Mountains in N. Ireland is a small area (30 ha) of relic woodland (Figure 4.2). The main species in the canopy, oak (*Quercus petraea*), is often found in association with birch (*Betula pubescens*). Other common tree species are hazel (*Corylus avellana*), rowan (*Sorbus aucuparia*) and ash (*Fraxinus excelsior*). Both the canopy and ground vegetation are heterogeneous, reflecting variations in slope, soil, geology and past management (Magurran, 1981, 1985).

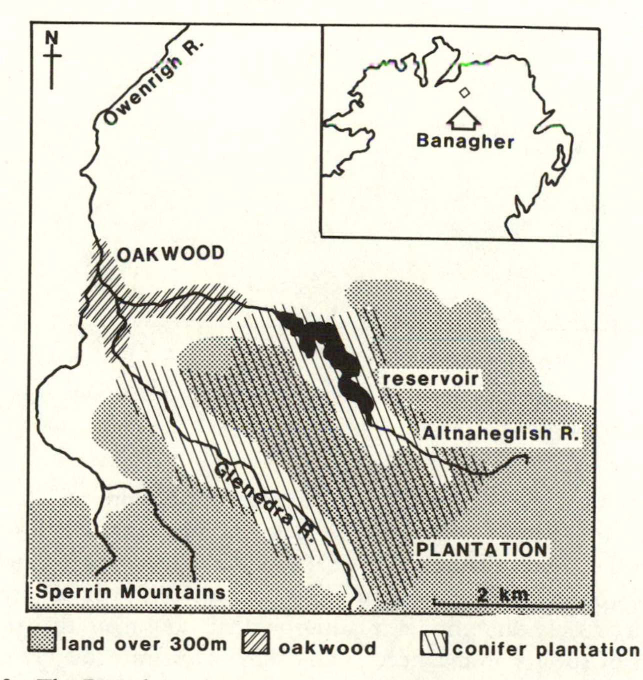

Figure 4.2 The Banagher oakwood and conifer plantation.

Adjacent to Banagher oakwood is a recently established and typical conifer plantation (Figure 4.2) covering an area of over 1000 ha. Pure stands of sitka spruce (*Picea sitchensis*) extend across 72% of the plantation. Over 85% of the planting took place between 1945 and 1965, much of it concentrated in two short periods, 1946–9 and 1960–4.

The variety and abundance of species of ground flora in the two woodlands was measured using randomly sited point quadrats (Chapter 3 and see Magurran, 1981, for details) while moth diversity was assessed by means of portable light traps (Magurran, 1985).

Rank abundance plots of the ground flora (Figure 4.3) show that the relic oakwood has more species and less dominance. The selection of diversity

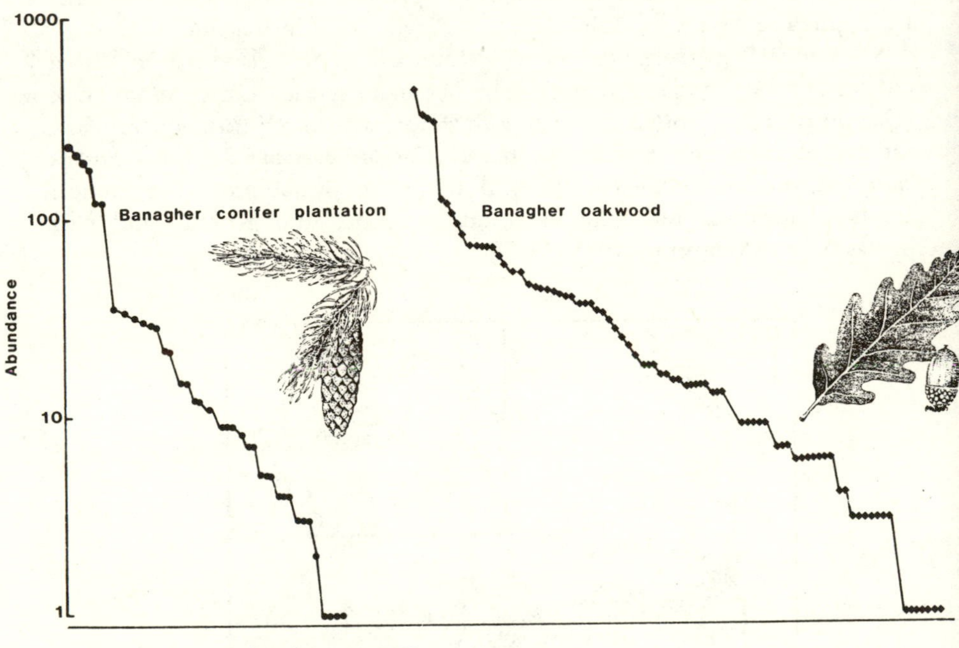

Figure 4.3 Rank abundance plots of ground vegetation in Banagher oakwood and Banagher conifer plantation.

indices listed in Table 4.3 confirms this first impression: in all cases the oakwood is considerably more diverse than the plantation. With so much evidence for the greater richness of the oakwood is it necessary to complete the laborious task of fitting the various models? The answer, perhaps surprisingly, is yes for it is only the conifer plantation that conforms to a log series distribution of species abundances (Table 4.3). This immediately arouses a suspicion that one factor is important in determining the number and

Table 4.3 Diversity of the Banagher woodlands. (A) Diversity indices and (B) fit of models.

	Oakwood	*Conifer plantation*
(A) Diversity		
Species richness (S)	86	45
Individuals (N)	3666	1543
Margalef	10.44	4.96
Berger–Parker (N_∞)	8.43	6.51
Simpson ($1/D$)	19.67	12.04
Shannon	3.54	2.90
Shannon evenness	0.80	0.76
Log series (α)	16.15	8.60
Log normal (λ)	129	68
(B) Fit of models		
Log series	No	Yes
Log normal	Yes	Yes
Broken stick	No	No

abundance of species of vegetation in the plantation. Not surprisingly it is the amount of light penetrating the canopy in the spring which emerges as the critical factor (Magurran, 1981). By contrast the only distribution which describes the oakwood is the truncated log normal. This suggests that the diversity of the vegetation in the deciduous woodland is subject to a range of influences. Support for this hypothesis comes from evidence that a combination of soil pH, light, slope, degree of waterlogging and disturbance is important in determining the level of species richness.

Large differences between the oakwood and the plantation are also revealed by the moth data (Magurran, 1985). Here again the rank abundance plots (Figure 4.4) show the oakwood to have a greater range of species. When the three models are formally tested the pattern is identical to that observed with the ground flora (Figure 4.5). The log series is appropriate only to the plantation while the truncated log normal fits both sites. The broken stick is a poor fit to the data because for both habitats it predicts fewer rare species than were recorded. Conversely, the log series provides an unsatisfactory fit to the oakwood data because it predicts too many rare species. All diversity indices show that the moth fauna of the oakwood is the most diverse. For example, λ, the log normal index, estimates the diversity of the oakwood as $\lambda = 163.3$ and the diversity of the plantation as $\lambda = 97.4$. The correlation with the diversity of the vegetation is striking. Many caterpillars have specific food requirements and it may be that the low diversity of the vegetation in the plantation is

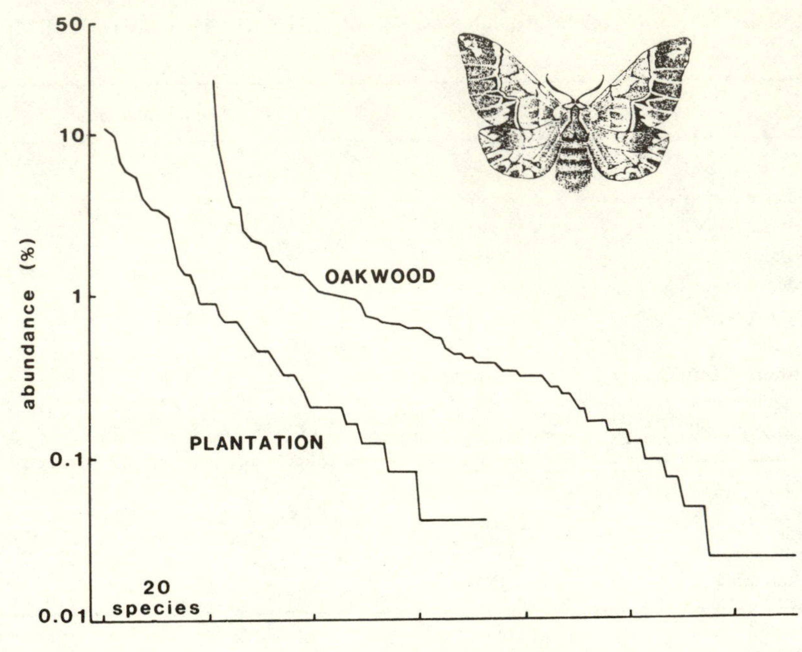

Figure 4.4 Rank abundance plots of moths in Banagher oakwood and Banagher conifer plantation.

limiting the moth diversity. However without further evidence it is perhaps best not to translate the observed correlation into a direct causation.

Every index tested, from log normal λ to the Margalef index, and from measures of richness to those of evenness, showed that Banagher oakwood was substantially more diverse than Banagher plantation. Yapp's assertion that diversity indices provide a measure of the deleterious effect of conifer plantations on wildlife is therefore vindicated. [It should be noted that Yapp's paper contains a number of statistical errors, particularly with regard to fitting the log series model. For details and corrections see Usher (1983).] But this type of comparison tells us little about the relative merits of the various diversity measures. Indeed differences as great as those beween Banagher oakwood and Banagher plantation would be detected by virtually any index that the ecologist cared to adopt or devise.

Goodness of fit tests

Tables 4.2 and 4.3 illustrate a common phenomenon in the measurement of diversity. Many sets of species abundance data are described by the truncated

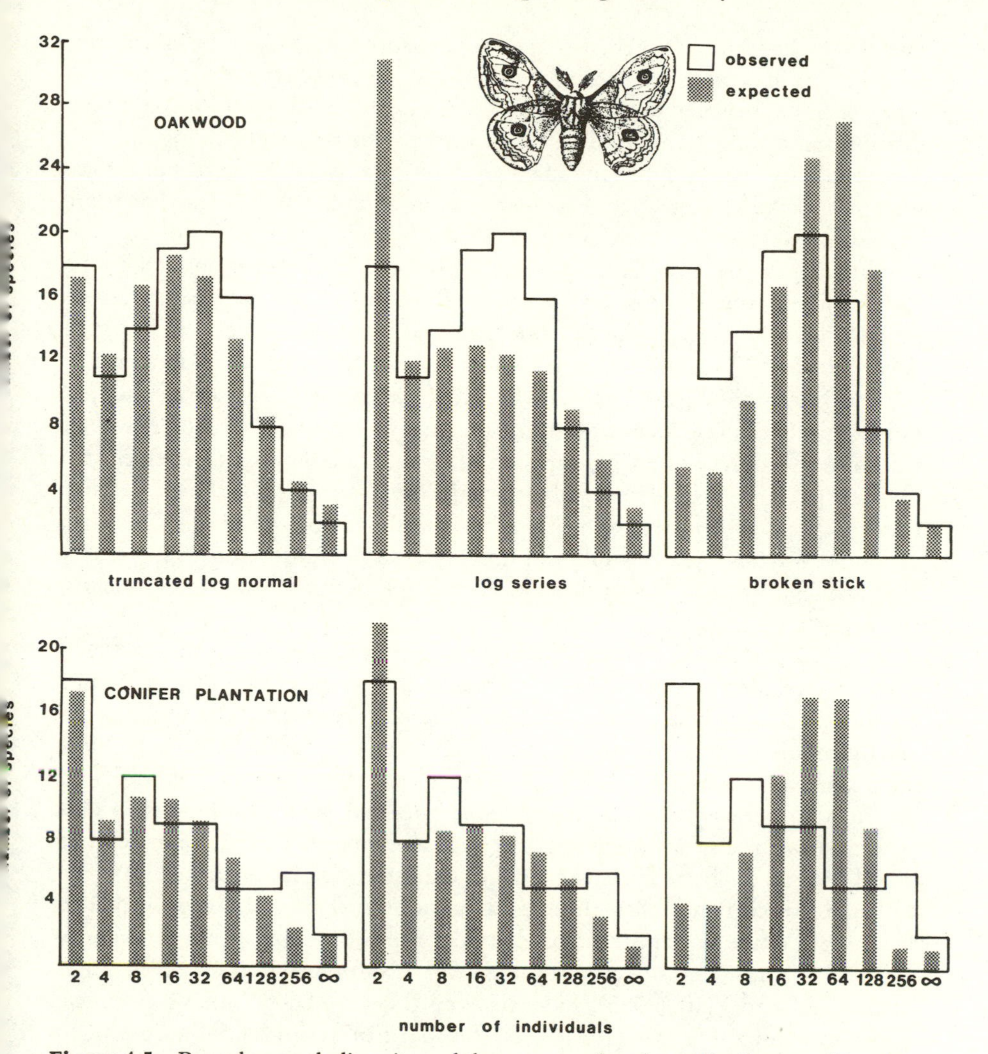

Figure 4.5 Banagher moth diversity and three species abundance distributions. Here the number of species observed in nine abundance classes (or octaves) is plotted against the number of species predicted by the log series, (truncated) log normal and broken stick models. Where there is a good agreement between the observed and expected a low value (non-significant) of χ^2 will result.

log normal and another model. In this case both the log series and the truncated log normal are appropriate fits to the data from the plantation. The explanation, discussed in detail in Chapter 2, is that the shape of a truncated log normal is not fixed. An almost fully veiled log normal resembles a log series (Figure 2.13). As the distribution is progressively unveiled, first the mode and eventually the symmetrical bell-shape become apparent. When the sample size is large, or has been determined using the diversity curve described in Chapter 3, it is reasonable to treat an observed log series distribution as the true distribution and not as a sampling distribution of an as yet unseen log normal. Similar advice would apply to a data set described by both the truncated log normal and the broken stick. Likewise species abundance data (such as those from the oakwood) which were fitted by the truncated log normal and nothing else can be treated as log normal. The problem is compounded by the fact that the goodness of fit tests are carried out on a small number of classes (usually less than 10) and that the differences between the models can lie in the way they allocate species between two or three of these classes. The whole χ^2 distribution can of course be used when comparing the fit of various models. For example if goodness of fit tests gave values of $\chi^2 = 10.7$ (with 6 degrees of freedom) for the truncated log normal and $\chi^2 = 2.1$ (with 7 degrees of freedom) for the log series it would be possible to make the statement that the probability of the expected truncated log normal being different from the observed data was $<90\%$ while the probability of the log series being different was $<10\%$. Both values are below the conventional level of 95% but the log series is clearly the much better fit.

Visual inspection of a graph showing the differences between the observed and expected species abundances is an invaluable way of interpreting the results of goodness of fit tests. For instance, Figure 4.5 shows that the log series is clearly predicting too many rare species and too few species of intermediate abundance in the conifer plantation. Very small data sets may sometimes be described by the log series, the truncated log normal *and* the broken stick. This is because with only a few species in each abundance class it can be difficult to detect differences between observed and expected distributions.

Much criticism has been directed at goodness of fit tests because of this failure to provide a clear distinction between the competing species abundance models. A number of investigators, for example Hughes (1986) and Lambshead and Platt (1985), have rejected their use in favour of graphical inspection alone. Hughes (1986) used the shape of a rank abundance plot to assess whether the log series or the dynamics model was the best predictor of species abundance patterns of 222 communities (see page 31) while the fit of the log normal model was judged on the basis of the presence or absence of a mode in the species abundance distribution. A glance at the rank abundance plots scattered through this book (many of which were chosen because they are particularly good examples of the various models) however emphasizes the point already made in Chapter 2. That is that it can be difficult to discriminate between models on the basis of the shape of the rank abundance plot alone. Thus the best solution to the problem in almost all cases will be to interpret the results both in terms of goodness of fit tests and of the shape of the species abundance data.

The discriminant ability of diversity measures

To be really useful diversity indices must be capable of detecting subtle differences between sites. Taylor (1978) recognized that one of the more important tests of the effectiveness of a diversity statistic is how well it discriminates between sites or samples that are not unduly different. This attribute is vital because a major application of diversity measures is to gauge the effects of pollution or other environmental stress on a single community or to choose the best example out of a group of similar habitats for conservation purposes (Chapter 6). This section therefore explores the discriminant ability of diversity measures.

Taylor (1978) examined the discriminant ability of eight diversity measures by using analysis of variance to test for between-site variation in the total annual moth samples (replicated over 4 years) from nine environmentally stable sites in the Rothamsted Insect Survey. Of all the indices he tested Taylor found that α (from the log series) was by far the best discriminator. Next, in order, came H' (the Shannon index), S (species richness), λ (the log normal index), the reciprocal of the Simpson index, and biomass (log N). The two other parameters of the log normal (S^* and σ) were useless at discriminating between sites.

Subsequent studies have extended the number of indices compared. Kempton (1979) looked at the discriminant ability of the members of Hill's family (Figure 4.6). Once again the Rothamsted moth data were employed but on this occasion the sample size was increased to 14 sites each replicated over 7 years. Orders of a between 0 and 0.5 (where $N_0 = S =$ number of species and $N_1 = \exp H'$, the transformed Shannon index) provided the highest degree of discrimination. Measures of a at either end of the scale proved unsatisfactory.

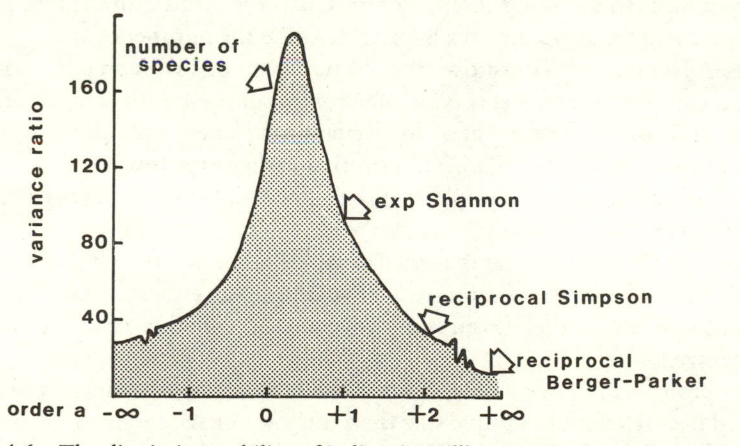

Figure 4.6 The discriminant ability of indices in Hill's series. Redrawn from Kempton (1979).

High orders of a (e.g. the Simpson and Berger–Parker indices) failed due to their dependence on the abundances of the most common species while low orders of a were unduly influenced by rare species. Kempton and Taylor (1976) showed that the degree of discrimination was greater for the transformed versions of the Shannon and Simpson indices (exp H' and $1/D$) than for their untransformed counterparts. Kempton and Wedderburn (1978) found that α and Q afforded a greater degree of discrimination than any form of the Shannon and Simpson indices.

In an investigation of the diversity of moths at ten light trap sites spread across Banagher plantation and Banagher oakwood Magurran (1981) found that the Margalef, McIntosh U and species richness S measures gave the greatest degree of discrimination. The Brillouin index emerged as better than both H' and exp H'. Evenness and dominance measures, for example Berger–Parker, Simpson, McIntosh D and Shannon and Brillouin evenness, were least sensitive to the differences between the sites. Morris and Lakhani (1979) similarly reported the Simpson index to be less sensitive to inter-site differences than the Shannon index.

Although the studies summarized above all tested slightly different sets of measures the general conclusion is that the indices weighted towards species richness are more useful for detecting differences between sites than the indices which emphasize the dominance/evenness component of diversity.

Sensitivity to sample size

Independence from sample size is a criterion frequently used to judge the effectiveness of diversity statistics. Species richness (S) is an index which is clearly subject to sampling intensity and Chapters 2 and 3 illustrated how S will increase as the sampling area is extended or as the number of samples taken increases. Kempton (1979) shows that S may even be biased in circumstances where a complete species list is available. For instance the annual totals of moth species at two woodland sites in Britain displayed considerable yearly fluctuations due to erratic changes in population densities. It was only when the results were corrected for sample size that consistent differences between the two sites were obtained.

Kempton (1979) looked at the small sample bias of diversity indices in Hill's (1973) series. The general finding was that the measures best at discriminating between sites (that is the measures of orders $a < 2$) were those most sensitive to sample size.

Kempton and Taylor (1974) found that the log series index α was less affected by variations in sample size than the log normal index. This attribute of α is a result of its dependence on the numbers of species of intermediate abundance – it is relatively unaffected by either rare species or common ones.

Similarly the Q statistic, which by its very nature describes species abundances in the inter-quartile region of the species abundance distribution, is robust against variable sample size. Kempton and Wedderburn (1978) estimated that Q will be unbiased when more than 50% of all species present appear in the sample while Taylor (1978) showed that α is completely independent of sample size if $N > 1000$. Taylor (1978) also demonstrated that the Simpson and Shannon indices are more sensitive to sample size than α.

Although evenness measures are not readily associated with small sample bias Peet (1974) showed that they can be seriously affected by sampling variations and concludes that estimates of evenness are only valid in circumstances where total species richness is known.

The Berger–Parker index is independent of species richness but is subject to bias caused by fluctuations in the abundance of the commonest species. A large sample size will help ensure that the true abundance of this species has been recorded, especially in situations where the individuals are aggregated rather than randomly dispersed.

Kempton and Wedderburn (1978) have concluded that absence of small sample bias should not be taken as the most important criterion when selecting a diversity index since even in the best circumstances small samples permit only a crude comparison between communities. The construction of a diversity curve (Chapter 3) helps ensure that sample size is adequate for the diversity index being used.

What aspect of diversity is the index measuring?

As Goodman (1975) observed, and as the graphs in Figure 4.7 confirm, diversity indices are often correlated. Magurran (1981) looked at this phenomenon in more detail by testing the concordance of rankings of sites when their diversities had been calculated using a variety of indices. Once again the Banagher moth data were used. The results are displayed in the triangular matrix in Table 4.4. The model based indices, α and λ, the Q statistic, species richness S, the information theory measures, and the Margalef and McIntosh U indices all produced significantly concordant rankings of sites. The indices that reflect dominance, that is Simpson, Berger–Parker and the Shannon, McIntosh and Brillouin evenness measures gave a different but also consistent ranking of sites.

Peet (1974) suggested that heterogeneity measures (the statistics that combine S and N) could be divided into Type 1 and Type 2 indices. Type 1 indices are those most affected by rare species (that is species richness) while Type 2 indices are sensitive to changes in the abundance of the commonest species (that is dominance). The best known examples of Type 1 and Type 2 measures are respectively the Shannon and Simpson indices.

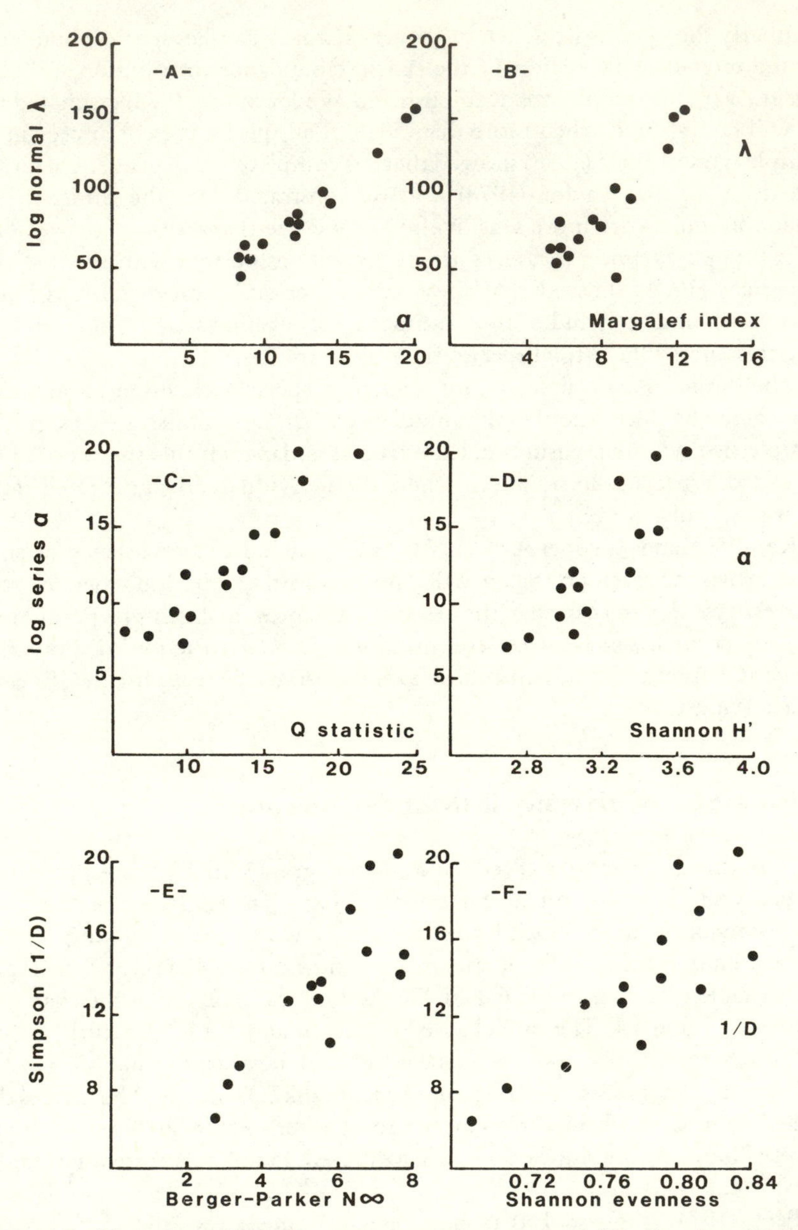

Figure 4.7 The values of diversity measures are often correlated. The diversity of 12 light-trapping sites in the Banagher woodlands was estimated using a range of diversity measures. The graphs are as follows: (A) log normal λ, and log series α; (B) log normal λ, and Margalef index; (C) log series α and Q statistic; (D) log series α and Shannon H'; (E) reciprocal Simpson and reciprocal Berger–Parker; (F) reciprocal Simpson and Shannon evenness.

Table 4.4 A comparison of diversity measures. The diversity of moths in ten areas in the Banagher woodlands was estimated using a range of diversity statistics. For each diversity index the sites were ranked from 1 to 10, that is from highest to lowest diversity. The concordance of rankings between pairs of indices was calculated using the Spearman rank correlation coefficient (r_s). Significant correlations are shown as ** ($P<0.01$) and * ($P<0.05$) while ns = not significant. Two groups of indices are present. The richness-based measures, for example S, α and Shannon give a concordant ranking of sites, while the dominance and evenness indices give a different but also consistent ranking.

	N	λ	α	Q	H'	HB	D_{Mg}	McU	$1/D$	N_∞	McD	$H'E$	HBE
S	**	**	**	**	*	**	**	**	ns	ns	ns	ns	ns
N		**	**	**	*	**	**	**	ns	ns	ns	ns	ns
λ			**	**	**	**	**	**	ns	ns	ns	ns	ns
α				**	*	**	**	**	ns	ns	ns	ns	ns
Q					*	**	**	**	ns	ns	ns	ns	ns
H'						**	*	*	ns	ns	ns	ns	ns
HB							**	**	ns	ns	ns	ns	ns
D_{Mg}								**	ns	ns	ns	ns	ns
McU									ns	ns	ns	ns	ns
$1/D$										**	**	**	**
N_∞											**	**	**
McD												**	**
$H'E$													**

S = number of species; N = number of individuals; λ = log normal index; α = log series index; Q = Q statistic; H' = Shannon index; HB = Brillouin index; D_{Mg} = Margalef index; McU = McIntosh U index; $1/D$ = reciprocal of Simpson's index; N_∞ = Berger–Parker index; McD = McIntosh dominance index; $H'E$ = Shannon evenness index; HBE = Brillouin evenness index.

Kempton (1979) noted that different diversity indices often produced inconsistent orderings of communities. He did however conclude that this inconsistency is rarer in field data than analyses using artificial and unrealistic data suggest. The discussion above supports this finding provided that indices from within either the species richness group or the dominance/evenness group are chosen.

Which measures are widely used?

Taken overall, species richness (S) is the most widely adopted diversity index. However the vogue for using measures incorporating species abundances has led to the widespread use of the Shannon index. Also fashionable is the Simpson index. The work of Taylor and his colleagues has encouraged the adoption of log series α and it is now the most popular of the parametric indices. Log normal λ, and the Q statistic, while having much to recommend them, are only infrequently applied. Also rare in this distribution of usage of diversity indices are the Margalef, McIntosh and Brillouin measures. With May's support the Berger–Parker index shows strong signs of being more frequently adopted.

Statistical tests

When diversity indices have been calculated a frequent response is that 'OK we now know that community A is more diverse than community B, but what does that really mean?'. In part this disenchanted reaction is because it is rare to attach statistical significance to differences in diversity. So the ecologist finding that the diversity (calculated using the Shannon index) of the bird fauna in two woodlands is $H' = 2.31$ and $H' = 1.95$ is left wondering whether the woodlands are really quite similar in terms of diversity or are in fact very different.

The initial answer to this question where the Shannon index is concerned is to calculate the variance and do a t test in the manner prescribed by Hutcheson (1970, and see Chapter 2). But these calculations are very tedious and in any case, for the reasons given below the Shannon index is not the best choice of diversity statistic.

A more satisfactory route can be followed in cases where replicate samples have been taken from the sites or communities to be compared. Repeated estimates of diversity are usually normally distributed (see Figure 4.8). This was the property that allowed Taylor and others to investigate the discriminatory ability of diversity measures. It also means that analysis of variance can be used to test for significant differences in the diversity of sites. For instance Gaudreault *et al.* (1986) used this technique to show that there were no significant differences between months in the diversity of the diets of sticklebacks (*Pungitius pungitius*) and juvenile brook charr (*Salvelinus fontinalis*) in Quebec. Full details of analysis of variance and of methods of transforming data that are not normally distributed are given by Sokal and Rohlf (1981).

Alternatively the jack-knife technique (see Chapter 2) can be used to improve the estimate of a diversity statistic, to obtain the standard error of the estimate and to attach confidence limits.

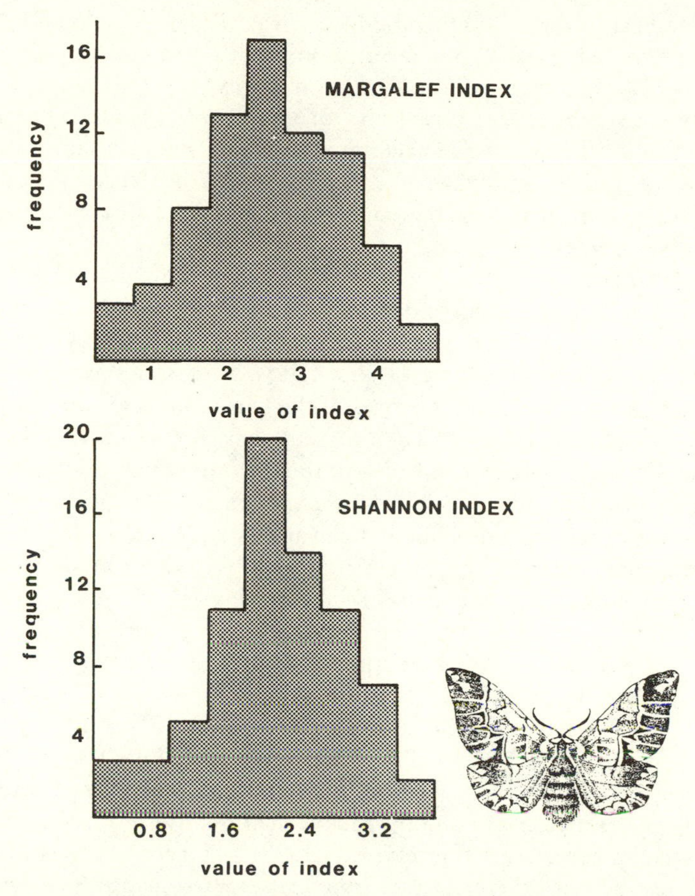

Figure 4.8 Repeated estimates of diversity from the same site are often normally distributed. This graph shows the distribution of values of the Margalef index and Shannon index calculated for light-trap catches in the Banagher conifer plantation.

Choice of index

There is little concensus on the best diversity measure to use and no index has received the backing of even the majority of workers in the field. The Shannon index in particular has attracted much criticism. May (1975) discussed its performance in relation to the broken stick, log normal and log series models and showed it to be a very insensitive measure of the character of the species abundance distribution. In place of the Shannon index May opted for the Simpson and Berger–Parker measures though he stressed that the full species abundance distribution should be examined wherever possible. Goodman (1975) similarly concluded that the Shannon was a 'dubious index' with 'no direct biological interpretation'. Like May, Peet (1974), Alatalo and Alatalo (1977) and Routledge (1979) prefer Simpson's index over the Shannon index. Pielou's (1975) advocacy of the Brillouin index has not resulted in its widespread use. Peet (1974) rejects the Brillouin index because it can give misleading answers. For instance in certain (fairly contrived) circumstances the Brillouin index may imply that a sample with the largest number of individuals (N) is more diverse than one with the greatest species richness and evenness (Peet, 1974).

Taylor (1978) came out strongly in favour of α, the log series index, because of its good discriminant ability and the fact that it is not unduly influenced by sample size. He also felt that α is a satisfactory measure of diversity, even when the underlying species abundances do not follow a log series distribution and that α is less affected by the abundances of the commonest species than either the Shannon or Simpson index. The only disadvantage of α is that it is based purely on S (species richness) and N (number of individuals). Thus α cannot discriminate situations where S and N remain constant, but where there is a change in evenness (such as in Hidden Glen and Wild Wood in Tables 4.2 and 4.3). But this again is largely an academic question as it is very unlikely that any genuine data collections will behave in this way. Furthermore the large number of investigations into the behaviour of α and its satisfactory performance in a wide range of circumstances make it an excellent candidate for a universal diversity statistic (Southwood, 1978).

The Q statistic has received considerable support from Taylor (1978), Kempton and Taylor (1976, 1978) and Kempton and Wedderburn (1978).

Yet despite these and other analyses the selection of diversity statistics has remained more a matter of fashion or habit than of any rigorous appraisal of their relative qualities. As Southwood (1978) observed 'there can be no universal best buy but there are rich opportunities for inappropriate usages'. On practical grounds it would be helpful if ecologists could standardize on the use of one or a few diversity statistics. This at least would make different data sets more comparable.

Table 4.5 summarizes the conclusions about the effectiveness of a range of diversity indices. Since the precise way in which a test of the performance of a diversity statistic is formulated will affect the conclusions drawn, this table should not be taken as an indication of how an index will respond in all circumstances. (For instance the sensitivity of a statistic to sample may vary according to whether the underlying pattern of species abundances is geometric series or log normal.) Instead it is a guide to the way in which the diversity measures will behave with realistic data from a number of genuine communities.

Guidelines for the analysis of diversity data follows. They are derived from the discussion above and also take into account many of the recommendations made by Southwood (1978).

Table 4.5 A summary of the performance and characteristics of a range of diversity statistics. As noted in the text these assessments are partly subjective and valid only when the statistics are applied to genuine data sets as opposed to highly artificial ones. The intention of the table is not to give a definitive classification of diversity measures but rather to show their relative merits and shortcomings. The simplicity or complexity of a calculation is judged from the viewpoint of a student with minimal mathematical experience and the most basic of pocket calculators. The evenness and dominance measures marked as simple* to calculate assume that the main index on which they are based has already been calculated. The column headed richness shows whether an index is biased towards either species richness on the one hand, or evenness (or dominance) on the other.

	Discriminant ability	*Sensitivity to sample size*	*Richness or evenness dominance*	*Calculation*	*Widely used?*
α (log series)	Good	Low	Richness	Simple	Yes
λ (log normal)	Good	Moderate	Richness	Complex	No
Q statistic	Good	Low	Richness	Complex	No
S (species richness)	Good	High	Richness	Simple	Yes
Margalef index	Good	High	Richness	Simple	No
Shannon index	Moderate	Moderate	Richness	Intermediate	Yes
Brillouin index	Moderate	Moderate	Richness	Complex	No
McIntosh *U* index	Good	Moderate	Richness	Intermediate	No
Simpson index	Moderate	Low	Dominance	Intermediate	Yes
Berger–Parker index	Poor	Low	Dominance	Simple	No
Shannon evenness	Poor	Moderate	Evenness	Simple*	No
Brillouin evenness	Poor	Moderate	Evenness	Complex	No
McIntosh *D* index	Poor	Moderate	Dominance	Simple*	No

1. Ensure where possible that sample sizes are equal and large enough to be representative (see Chapter 3 for advice).
2. Draw a rank abundance graph (see Chapter 2). This should provide a first indication as to whether the data follow the geometric series, log series, log normal or broken stick distributions.
3. Calculate the Margalef and Berger–Parker indices (see Chapter 2 for details). These straightforward measures give a quick measure of the species abundance and dominance components of diversity. Their ease of calculation and interpretation is an important advantage.
4. Determine log series α. This can be obtained by calculation (see Chapter 2)

or read directly from Williams's nomograph (Williams, 1964; Southwood, 1978). The work of Taylor and his colleagues provides strong support for the adoption of α as the standard diversity statistic. The Q statistic is a suitable alternative if α is felt to be inappropriate.

5. In studies in which diversity forms the most important theme it will often be valuable to test the fit of main species abundance models formally (see Chapter 2 for methods). This step is likely to be of most interest if the communities under investigation form a successional sequence or are subject to environmental stress. Interpret goodness of fit tests by referring to the rank abundance plots of 2 above and by inspecting graphs which superimpose observed and expected species abundance patterns (for example Figure 4.5).
6. When replicate samples have been taken use analysis of variance to test for significant differences between communities (see above).
7. The jack-knife procedure (Chapter 2) is a useful method of improving the estimate of a diversity statistic and attaching a confidence interval.
8. If one study is to be directly compared with another it is important to be consistent in choice of diversity index. For this reason it may be more informative to continue use of for example the Shannon index rather than switching to theoretically and biologically more acceptable indices.

Summary

The large number of diversity statistics available means that it may be difficult to select the most appropriate method of measuring diversity. When applied to realistic data sets these diversity indices can be divided into two categories. On one hand there are the indices which reflect the species richness element of diversity while on the other hand there are measures which express the degree of dominance (evenness) in the data. As a general observation, indices in the first category are better at discriminating between samples but are more affected by sample size than the dominance/evenness set of diversity measures. For reasons of standardization it would be prudent if ecologists would concentrate on one or a few indices. The log series index α, the Berger–Parker dominance index, and a measure of species richness (either S or the Margalef index) appear to combine most satisfactorily the advantages of being simple to calculate, easy to interpret and statistically and ecologically sound. In many cases it is valuable to go beyond a single diversity statistic and examine the shape of the species abundance distribution.

5
A variety of diversities

So far this book has concentrated on the measurement of species diversity. Yet there are many studies concerned with other varieties of diversity. Attempts by ecologists to explain why some areas are species rich and others are species poor or why a species is abundant in one location but rare in another often prompts an investigation of habitat diversity. In undertaking a study of habitat diversity ecologists are asking similar questions to the ones they pose when describing species diversity. The methods devised for measuring species diversity are also employed when niche width is being investigated. Niche width is, after all, a measure of the diversity of resources utilized. The first section of this chapter therefore looks at other contexts in which measures of species diversity can be utilized.

A rather different approach is required when ecologists wish to ascertain how species numbers and identities differ between communities or along gradients. Methods of describing this alternative variety of diversity, known as β (beta) or differentiation diversity, are reviewed in the second part of the chapter.

Structural and habitat diversity

At the simplest level habitat diversity is nothing more than the number of habitat types in a defined geographical area. As such it is directly analogous to species richness, the most straightforward of the species diversity measures (Chapter 2). However before even this most basic assessment of habitat diversity can take place it is necessary to have a system of habitat classification.

Classification of habitats and structures

Elton and Miller (1954) pioneered the investigation of habitat diversity with their habitat classification scheme. This scheme operates at four levels. First the major habitat system (e.g. terrestrial or aquatic) is recognized. This habitat system is then allocated a formation type (e.g. woodland or open ground) and the presence of vertical layers (e.g. ground flora, shrub, high canopy) is recorded along with qualifiers (e.g. conifer, deciduous). The classification scheme was designed for use with punch cards and did much to encourage the quantitative recording of habitat diversity in the days before computers were widely used by ecologists. It was employed (with slight modification) by the British Nature Conservancy in the 1950s and 60s in the ecological assessment of natural and semi-natural areas. In recent years many more schemes for recording habitat diversity have been devised (Kirby *et al.*, 1986, list a selection) and habitat diversity has become established as an important component of wildlife conservation evaluation (Pearsall *et al.*, 1986; Fuller and Langslow, 1986; Usher, 1986). Often these schemes use an index of habitat diversity similar to the indices of species diversity described in Chapter 2.

Elton (1966) was primarily concerned with woodland ecology and it is therefore appropriate that this is the 'formation type' in which some of the most interesting work on habitat diversity has been carried out. The concept of structural diversity, that is the number of vertical layers present and the abundance of vegetation within them, has proved important in studies of the diversity of woodland bird communities. In a classic paper MacArthur and MacArthur (1961) found that the structural diversity of temperate woodlands in North America was a much better predictor of bird species diversity than was plant species diversity (Figure 5.1). Correlations between bird species diversity and woodland structural diversity (commonly referred to as foliage height diversity) have also been recorded in Central America (MacArthur *et al.*, 1966; Karr and Roth, 1971), Australia (Recher, 1969) and Europe (Moss, 1978).

MacArthur and MacArthur (1961) obtained foliage height diversity by

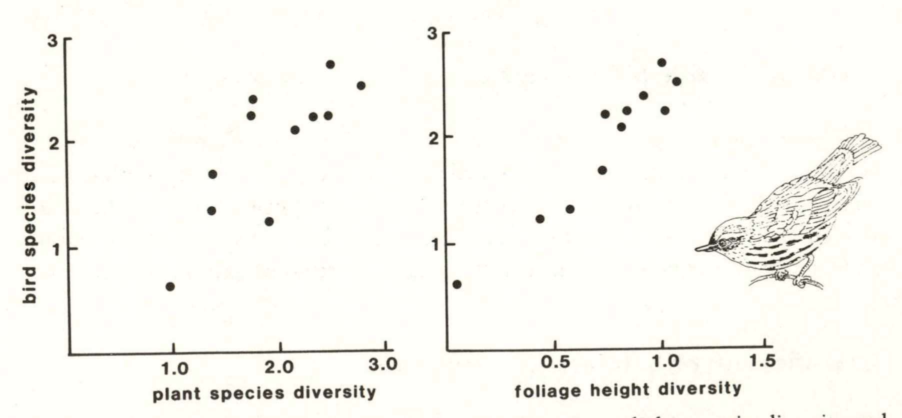

Figure 5.1 The relationship between bird species diversity and plant species diversity and structural diversity (foliage height diversity) in deciduous forest plots in the eastern United States. Redrawn from MacArthur and MacArthur (1961).

visually estimating the proportion of total foliage in chosen horizontal layers. They found that the best relationship between bird species diversity and foliage height diversity (with both diversities calculated using the Shannon index) was obtained using three horizontal layers (0–0.7 m, 0.7–7.6 m and >7.6 m) of vegetation. A similar procedure was adopted by MacArthur and Horn (1969), Terborgh (1977) and Moss (1978). Blondel and Cuvillier's (1977) stratiscope facilitates the measurement of structural diversity in woodlands.

In their investigation of the relationships between plant and insect diversities during a young field to woodland succession in southern England, Southwood *et al.* (1979) chose to divide structural diversity into two components. First they estimated plant spatial diversity by recording the number of touches by the vegetation to a vertical pin or pole. This allowed them to construct spatial diversity profiles for the three phases of the succession (Figure 5.2). Then they measured architectural complexity which was defined as the number of categories of architecture into which the plant structure in each site could be divided (Table 5.1). The diversity of both forms of structural diversity, as well

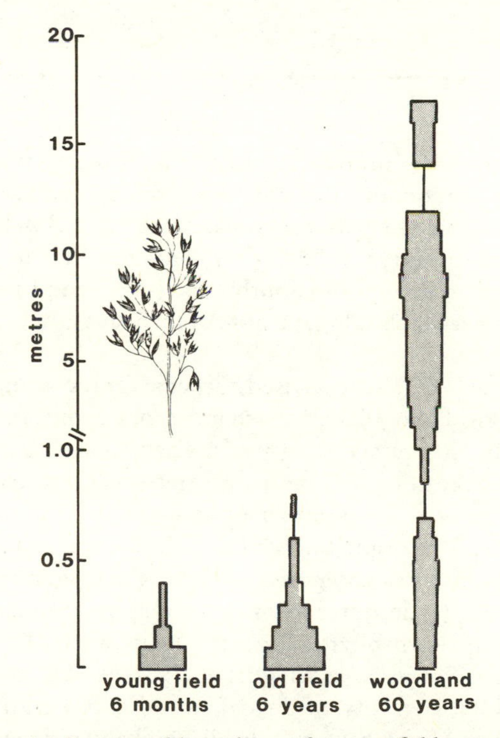

Figure 5.2 The spatial diversity of three phases of a young field to woodland succession in southern England. These profiles show the vertical distribution of vegetation. In the woodland the canopy is multilayered while in the old field stratification is just beginning. Redrawn from Southwood *et al.* (1979).

Table 5.1 The categories of architectural complexity used by Southwood *et al.* (1979). Although these categories are defined in botanical terms they also reflect the types of the micro-habitats occupied by invertebrates.

Dead wood >10 cm diam.	Flower buds
Dead wood >2 cm and <10 cm diam.	Open flowers
Dead wood <2 cm diam.	Dead flowers
Bark on dead wood >10 cm	Ripening/ripe fruits (seeds)
Bark on dead wood 2–10 cm	Old fruiting structures
Bark on dead wood <2 cm	Dead stems
Bark on living wood >10 cm	Dead leaves
Bark on living wood 2–10 cm	Mosses – epiphytes
Bark on living wood <2 cm	Mosses – on soil surface
Green stems	Liverworts – epiphytes
Leaves of monocotyledons	Liverworts – on soil surface
Petioles	Lichens and algae – epiphytes
Leaf surface – upper	Lichens and algae – on soil
Leaf surface – lower	Fungal fruiting bodies – on soil
Leaf/buds/scales	Fungal fruiting bodies – on vegetation
Flowering stems	

as of plant and insect taxonomic diversity, was estimated using the log series diversity index, α. Taxonomic diversity was also expressed as species richness. The results showed that insect diversity was much more closely related to plant architectural diversity and spatial diversity combined than to plant taxonomic diversity (Figure 5.3). Brown and Southwood (1987) emphasize that measures of architectural diversity should take note of the ways in which insects exploit plant structures.

Bunce and Shaw (1973) have devised a scheme for recording the diversity of habitats within British woodlands which incorporates the taxonomic diversity of the trees and the structural diversity of the habitat as well as its architectural complexity. The recording scheme is integrated with a standard ecological study and has been used in a number of major woodland surveys in Britain (Spellerberg, 1981). Bunce and Shaw's list gives 82 types of habitat subdivided into seven categories. These categories are (a) tree management; (b) the species of tree regenerating; (c) dead tree habitats; (d) epiphytes and lianes on trees; (e) rock habitats; (f) aquatic habitats; (g) open habitats. Such a list is easy to use and simple to interpret. For instance Magurran (1981) used a modified version to compare the habitat diversity of an oak wood and a conifer plantation at Banagher, N. Ireland (see Chapter 4). In all cases habitat diversity was lower in the conifer plantation (Figure 5.4). Nevertheless there were interesting differences between the stand types within the plantation with the deciduous larch emerging as the most diverse.

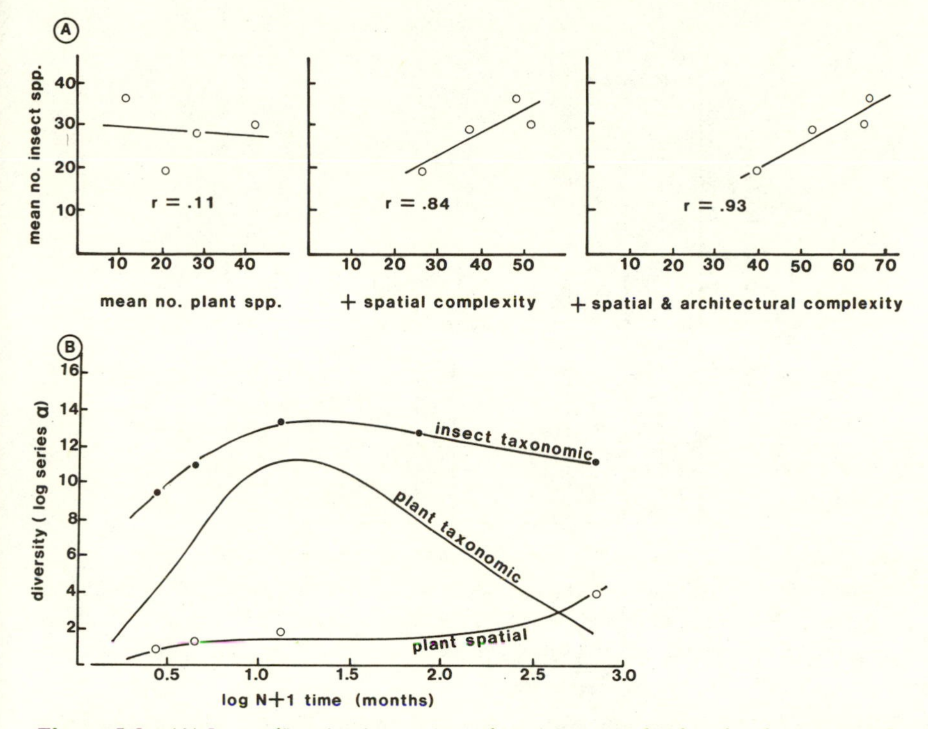

Figure 5.3 (A) Insect diversity (as species richness) is more closely related to structural diversity (especially as spatial and architectural complexity combined) than to plant species diversity. (B) The levels of diversity (insect taxonomic, plant taxonomic and plant spatial) change during the course of the succession. Diversity in this instance is measured using log series α. Redrawn from Southwood *et al.* (1979).

One method of measuring canopy structure involves the use of hemispherical photography. Although originally devised for the study of light conditions in woodlands (Hill, 1924; Anderson, 1964, 1971), this technique can be adapted to provide detailed information on the density and distribution of foliage. The canopy is photographed using a 180° fish-eye lens (Evans *et al.*, 1975; Pope and Lloyd, 1975). By superimposing a grid which divides the photographs into 1000 sections, each of which accounts for 0.01% of the total irradiance reaching the ground (Anderson, 1964), it is possible to measure the percentage canopy cover accurately. More detailed information can be obtained by calculating the cover produced by different layers of vegetation. Canopy photographs taken in the Banagher woodlands are shown in Figure 5.5.

Canopy photography is only one way in which a fish-eye lens can be used to measure habitat structure. A novel approach was adopted by Burger (1972) who used this method to measure the structure of the vegetation, and the

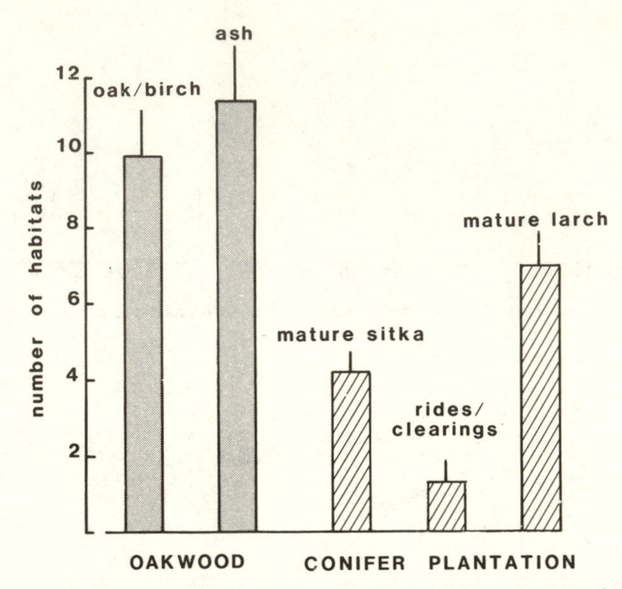

Figure 5.4 Habitat diversity. The mean (and 95% confidence limits) of the number of habitats in areas of oak/birch and ash within the Banagher oakwood, and stands of mature sitka spruce and mature larch, plus rides and clearings in the Banagher conifer plantation.

degree of cover, around nests of the Franklin Gull (*Larus pipixcon*) in Minnesota, USA.

Measures of habitat diversity are of course not restricted to terrestrial environments. The number of substrate types has been shown to be a good predictor of species diversity for marine decapod insects (Abele, 1974), freshwater molluscs (Harman, 1972) and benthic invertebrates (Allan, 1975). All these organisms spend their adult life in the substrate so the relationship between substrate diversity and species diversity is hardly surprising. For aquatic animals occupying a three-dimensional environment a more complex method of assessing structural diversity is required. Gorman and Karr (1978) concluded that they needed to take depth, current and bottom type into account when investigating the link between habitat diversity and the diversity of stream fish communities in Indiana and Panama (Figure 5.6), while Roberts and Ormond (1987) found that holes in coral were the best predictor of fish abundance in Red Sea fringing reefs.

Harper (1977) stresses the importance of taking an organism's eye-view of community diversity. This comment is as relevant to structural diversity as it is to species composition. For instance a quadrat that may appear species rich and structurally heterogeneous to the ecologist observing it may be perceived as homogeneous by the plant or insect living within it. Conversely an apparently unvarying strip of pasture may offer the grazing mollusc or sheep considerably

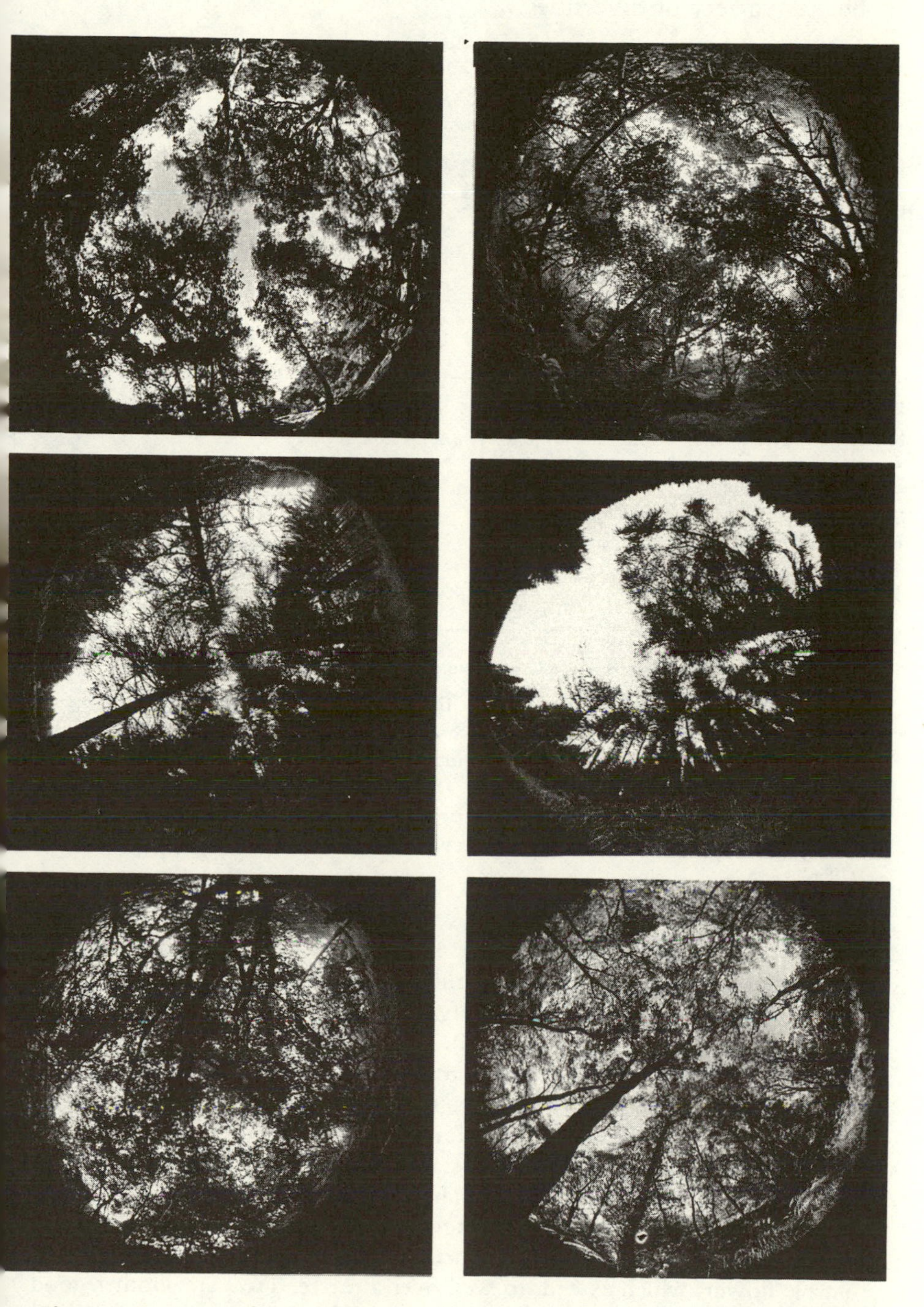

Figure 5.5 Hemispherical photographs (taken with a fish-eye lens) of the Banagher canopy.

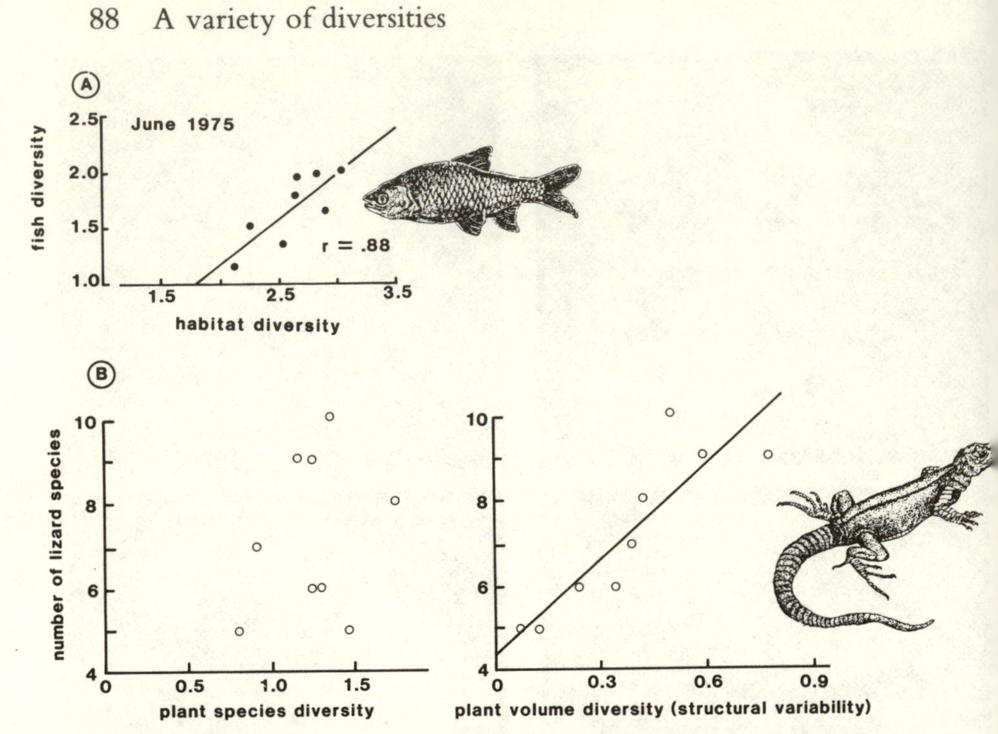

Figure 5.6 (A) The relationship between fish species diversity and habitat diversity in streams in Indiana and Panama. This graph shows the situation in June 1975. At other times of the year algal blooms and/or streams drying up may remove the significant correlation. Redrawn from Gorman and Karr (1978). (B) Like MacArthur and MacArthur's birds (Figure 5.1) the diversity of flatland lizard species in the southwestern United States is more closely related to structural diversity than to plant species diversity. Redrawn from Pianka (1966).

more variety than is at first apparent. Harper (1977) describes four forms of diversity that contribute to the structural diversity of plant communities.

First of all there is the somatic polymorphism of the parts of a genet (or functional individual – see Chapter 3). For instance the same plant may have different leaf forms on its juvenile and mature branches (for example *Eucalyptus* spp.), at different times of the year (for example desert shrubs) or on its flowering and non-flowering parts (for example *Valeriana dioica*).

Next comes the diversity of age-states within the community. Old and young plants of the same species often have markedly different growth forms. For instance in its first year the foxglove (*Digitalis purpurea*) is a prostrate rosette while by its second year it has acquired a spiral of leaves and a raceme of purple flowers which extends to well over a metre. Trees are another good example of plants which vary greatly in their growth form, and ecological role, at different phases of their life.

The third of Harper's categories concerns the genetic variants within a species. Referring to his own work on the white clover *Trifolium repens* he describes six genetic polymorphisms which can be found with a 1 m^2 area of grassland. These include the presence or absence of cyanogenic glycosides which affect palatability to slugs and other predators, genetic variation in leaf size, genetic variation in aggressiveness to grass species in the sward, and the occurrence of leaf marks which appear to be used as 'search images' by grazing sheep.

The diversity of microsites within the habitat is the final form of variety. Ridges and furrows in a permanent pasture each have characteristic species associated with them and differ considerably in their species richness. Variation in soil texture, drainage, exposure and countless other environmental factors can influence the identities and abundances of species found in a particular habitat.

Diversity measures

Once the habitat and structural types have been defined it is relatively simple to assess their diversity. Most studies opt for a simple species richness-type count of types but the Shannon index is also popular, especially in investigations of structural diversity. The work of Southwood and his colleagues has proved that α, the log series index, is a useful tool in the measurement of habitat diversity. The indices are calculated using the methods already described in Chapter 2.

Niche width

Niche width is a measure of the breadth or diversity of resources used by an individual or species. The usual approach is to use either the Shannon index (equation 2.17, page 35) or the Simpson index (equation 2.27, page 39) to calculate the width of the niche. The number of resource categories observed (for example, types of food eaten, varieties of habitat utilized, kinds of behaviour employed) replace number of species in the equation. Clearly a separate value must be calculated for each type of resource. Measures of abundance will depend on the way in which the index is being used. For instance if the niche width of a particular species is under consideration then abundance may be measured as the number of individuals either eating each type of food, living in each sort of habitat, or adopting each kind of behaviour. If, on the other hand, a measure of the niche width of an individual is required, then abundance can be taken as the amount of each food type eaten, the time spent in each habitat or the frequency with which each behaviour is performed.

An extensive literature deals with the measurement of niche width. Some of the more useful references include Colwell and Futuyma (1971), Feinsinger *et al.* (1981), Giller (1984), Hurlbert (1978), Southwood (1978) and Thormon (1982).

There are many examples of measures of niche width and such studies can contribute to the understanding of mechanisms involved in structuring communities. In one, Kotrschal and Thomson (1986) measured the trophic diversity, that is the width of the feeding niche, of 34 species of Pacific blennioid fish. The gut contents of the fish were identified and the abundances of over 70 categories of food type estimated. The trophic diversity of each species was then calculated using the Shannon index. These measures of trophic diversity were used to distinguish three categories of fish: (1) specialists (six species); (2) low diversity feeders (18 species); and (3) high diversity generalists (10 species). Kotrschal and Thomson found that the high diversity generalists, that is those species with wide feeding niches, were numerically more abundant than the low diversity feeders or specialists.

Caveats concerning measures of habitat diversity and niche width

Measures of niche width and habitat diversity are beset by the same problems which were encountered with indices of species diversity. Using a modification of the Shannon index to describe habitat diversity does not magically overcome the fact that this measure can be biased when sample sizes are small. Similarly, the Simpson index remains a dominance index whether it is used to measure the diversity of species or the diversity of resources. Problems will inevitably arise if sample sizes are too small or too variable. The advice offered with regard to the choice of species diversity indices is also relevant in this context. It is best to confine attention to a small number of indices whose properties are well known and which can be readily interpreted.

An additional hazard not encountered in measures of species diversity may confront the investigators of niche width and habitat diversity. Misidentifications and taxonomic quibbles apart, species are well defined entities. Thus species x and species y will always remain species x and species y even when they are recorded by different ecologists working in different continents. Classifications of habitat type and resource use are however often devised afresh for each study. Such unique classifications will usually preclude a direct comparison between investigations or even make it impossible (if insufficient information is available) for one worker to replicate another's study.

It is thus important to exercise care and a considerable degree of common sense when interpreting measures of these other types of diversity. Careful consideration must be given to the type of habitat or resource classification employed and it must be used consistently. Full details of classifications should be provided. It is vital that sample sizes should be consistent and large enough to represent the diversity adequately.

β or differentiation diversity

β diversity is essentially a measure of how different (or similar) a range of habitats or samples are in terms of the variety (and sometimes the abundances) of species found in them. One common approach to β diversity is to look at how species diversity changes along a gradient (Wilson and Mohler, 1983). Another way of viewing β diversity is to compare the species compositions of different communities. The fewer species that the different communities or gradient positions share, the higher the β diversity will be.

The term β diversity was coined by Whittaker (1960, 1977) whose four scales of inventory diversity (see Chapter 3 for details) are matched by three levels of differentiation diversity (pattern diversity, β diversity and delta diversity). β diversity is essentially the same as MacArthur's (1965) between habitat diversity. Delta diversity is defined as the change in species composition and abundance between areas of gamma diversity which occur within an area of epsilon diversity. It represents differentiation diversity over wide biogeographic areas. At the other end of the spectrum is pattern diversity which is conventionally defined as the differentiation diversity between samples taken from within a homogeneous habitat. β diversity is by far the most widely studied scale of differentiation diversity and indeed the term is often applied to any investigation which looks at the degree to which the species compositions of samples, habitats or communities differ (Southwood, 1978). Taken together with measures of within habitat diversity, β diversity can be used to give the overall diversity of an area (Routledge, 1977).

Wilson and Shmida (1984) have recently assessed six methods of measuring β diversity using presence and absence data. These are:

1. Whittaker's measure β_W

The first, and one of the most straightforward, measures of β diversity was introduced by Whittaker (1960).

$$\beta_W = S/\alpha - 1 \tag{5.1}$$

where S = the total number of species recorded in the system (i.e. gamma diversity) and α = the average sample diversity where each sample is a standard size and diversity is measured as species richness.

2. Cody's measure β_C

Cody (1975) was interested in the change in the composition of bird communities along habitat gradients. His index, which is easy to calculate and a good intuitive measure of species turnover, simply adds the number of new species encountered along a transect to the number of species which are lost.

$$\beta_C = \frac{g(H) + l(H)}{2} \tag{5.2}$$

where $g(H)$ = the number of species gained along the habitat transect and $l(H)$ are the number of species lost over the same transect.

3, 4 and 5. Routledge's measures, β_R, β_I and β_E

Routledge (1977) was concerned with how diversity measures can be partitioned into alpha and β components. The following three indices are derived from his work. The first measure, β_R, takes overall species richness and the degree of species overlap into consideration.

$$\beta_R = \frac{S^2}{(2r + S)} - 1 \tag{5.3}$$

where S = the total number of species in all samples and r = the number of species pairs with overlapping distributions.

β_I, the second index, stems from information theory and has been simplified for qualitative data and equal sample size by Wilson and Shmida (1984).

$$\beta_I = \log(T) - [(1/T)\Sigma e_i \log(e_i)] - [(1/T)\Sigma \alpha_j \log(\alpha_j)] \tag{5.4}$$

where e_i is the number of samples along the transect in which species i is present, α_j is the species richness of sample j and $T = \Sigma e_i = \Sigma \alpha_j$.

The third index β_E is simply the exponential form of β_I:

$$\beta_E = \exp(\beta_I) - 1 \tag{5.5}$$

6. Wilson and Shmida's measure, β_T

Wilson and Shmida (1984) proposed a sixth measure of β diversity. This index has the same elements of species loss (l) and gain (g) that are present in Cody's measure and the standardization by average sample richness α, which is a component of Whittaker's measure.

$$\beta_T = [g(H) + l(H)]/2\alpha \tag{5.6}$$

Worked examples of all six measures are shown in Example 14 (page 162).

Wilson and Shmida chose four criteria, number of community changes, additivity, independence from alpha diversity and independence from excessive sampling, to evaluate the six measures of β diversity. The degree to which each index measured community turnover was tested by calculating the β diversity for two hypothetical gradients, one of which was homogeneous, that is the same species were present throughout its length, and one which consisted of distinct communities with no overlap. Whittaker's index, β_W accurately reflected these extremes of community turnover. β_T was more limited in that it only adequately represented turnover in conditions where the alpha diversity at both ends of the gradient was equal to average alpha diversity. β_R and β_E were even more restricted in that they required constant species richness. The remaining two measures, β_C and β_I, showed no ability to pick up turnover.

The second criterion was additivity, that is the ability of a measure to give the same value of β diversity whether it is calculated using the two ends of a gradient or from the sum of the β diversities obtained within the gradient. For instance with three sampling points (a, b and c), β(a, c) should equal β(a, b) + β(b, c).

Only one index β_C was completely additive. When tested with field data, three of the remaining measures were found to be nearly additive with errors of 4% (β_T), 18% (β_W) and 24% (β_E).

Independence from alpha diversity, the third property, was examined by using β to compare two gradients which were identical except that one had twice as many species as the other. β_C alone failed this test. Without this independence it would be impossible to compare β diversity in species-rich and species-poor communities.

The final criterion, independence from sample size, was tested by increasing the number of (identical) samples taken at each site. All measures apart from the information-theory-derived β_I and β_E, were found to be unaffected by sampling in this restricted situation where all other information remained constant.

Out of the six measures β_W emerged as fulfilling most criteria with fewest restrictions. Wilson and Shmida's own index, β_T came a close second.

Wilson and Shmida (1984) tested the β diversity measures further by using them to examine vegetation communities along an altitudinal gradient on Mount Hermon in Israel. The presence and absence of species was recorded at 100 m intervals of altitude, commencing at 400 m above sea level. Four measures are plotted in Figure 5.7. Since values for $\beta_I \approx \beta_E$ the latter is excluded. β_C is not shown on the same graph because the unstandardized results are not directly comparable with those of the other measures. Interestingly, despite the diverse origins of the β diversity measures, the shapes of the curves are virtually identical. They all show the transition from maquis to montane vegetation which occurs between 1200 m and 1300 m and pick up the large shifts in the β diversity of the alpine flora above 2600 m.

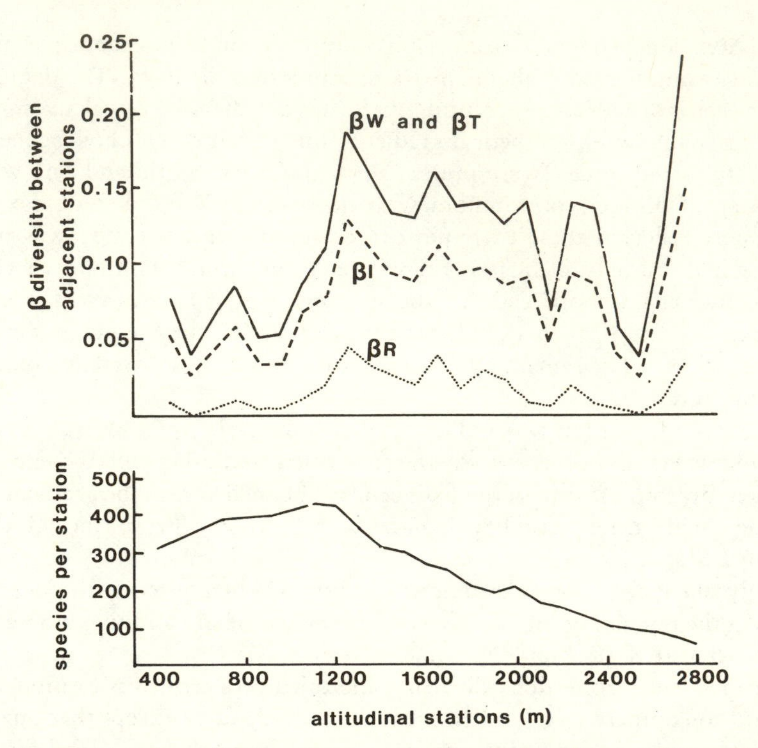

Figure 5.7 (A) Values of β diversity along an altitudinal gradient on Mount Hermon, Israel. β_W = Whittaker's measure, β_T = Wilson and Shmida's measure, while β_I and β_R are two of Routledge's measures. For more details see the text. (B) The number of species at each station along the gradient. Redrawn from Wilson and Shmida (1984).

All the above measures use presence and absence data. Guidelines for measuring β diversity with quantitative data are discussed by Wilson and Mohler (1983). Further techniques for the analysis of diversity patterns on environmental gradients are described by Pielou (1975) in Chapter 6 of her book *Ecological Diversity*.

Since β diversity is the variation in species composition between areas of alpha diversity there is no reason why it should be investigated only in terms of transects or environmental gradients. An alternative approach to the measurement of β diversity is to investigate the degree of association or similarity of sites or samples using standard ecological techniques of ordination and classification (Greig-Smith, 1983; Pielou, 1984; Southwood, 1978).

The easiest way to measure the β diversity of pairs of sites is by the use of similarity coefficients. A vast range of similarity indices exist (Clifford and Stephenson, 1975). However some of the oldest similarity coefficients are also the most useful. Particularly widely used are the Jaccard index and Sorensen index (Southwood, 1978; Janson and Vegelius, 1981, and see Example 15).

$$\text{Jaccard} \quad C_J = j/(a+b-j) \tag{5.7}$$

$$\text{Sorenson} \quad C_S = 2j/(a+b) \tag{5.8}$$

where j = the number of species found in both sites and a = the number of species in Site A with b the number of species in Site B. These indices are designed to equal 1 in cases of complete similarity (that is where the two sets of species are identical) and 0 if the sites are dissimilar and have no species in common. One of the great advantages of these measures is their simplicity. However this virtue is also a disadvantage in that the coefficients take no account of the abundances of species. All species count equally in the equation irrespective of whether they are abundant or rare. This consideration has led to similarity measures based on quantitative data (Southwood, 1978). Perhaps the most widely used is the version of the Sorensen index modified by Bray and Curtis (1957) (Example 15, page 165).

$$\text{Sorenson quantitative} \quad C_N = \frac{2_{jN}}{(aN+bN)} \tag{5.9}$$

where aN = the total number of individuals in site A, bN = the total number of individuals in site B, and, jN = the sum of the lower of the two abundances recorded for species found in both sites. Thus if 12 individuals of a species were found in Site A and 29 individuals of the same species in Site B the value 12 would be included in the summation to give jN.

Wolda (1981) investigated a range of quantitative similarity indices and found that all but one, the Morisita–Horn index (Worked Example 15), were strongly influenced by species richness and sample size. A disadvantage of the Morisita–Horn index however is that it is highly sensitive to the abundance of the most abundant species. Nevertheless Wolda (1983) successfully used a modified version of the Morista–Horn index to measure β diversity in tropical cockroaches.

$$\text{Morisita–Horn} \quad C_{mH} = \frac{2\Sigma(an_i bn_i)}{(da+db)aN \cdot bN} \tag{5.10}$$

where aN = total number of individuals in site A and an_i = number of individuals in the ith species in A.

$$da = \frac{\Sigma an_i^2}{aN^2}.$$

A recent extensive evaluation of similarity measures (Smith, 1986) tested both qualitative and quantitative techniques using data from the Rothamsted Insect Survey (Taylor, 1986). Smith concluded that the presence/absence (qualitative) were generally unsatisfactory. Of those tested the best proved to be the Sorensen index (equation 5.8). The large number of quantitative similarity measures made selection difficult and Smith advised that the choice of index for any particular study should depend on the form of the data and the aims of the investigation. She did however find (like Wolda, 1981) that versions of the Morisita–Horn index (equation 5.10) are among the most satisfactory available.

When there are a number of sites in the investigation a good representation of β diversity can be obtained through cluster analysis. Cluster analysis starts with a matrix giving the similarity between each pair of sites. The two most similar sites in this matrix are combined to form a single cluster. The analysis proceeds by successively clustering similar sites until all are combined in a single dendrogram (Figure 5.8). There are a variety of techniques for deciding how sites should be joined into clusters and how clusters should be combined

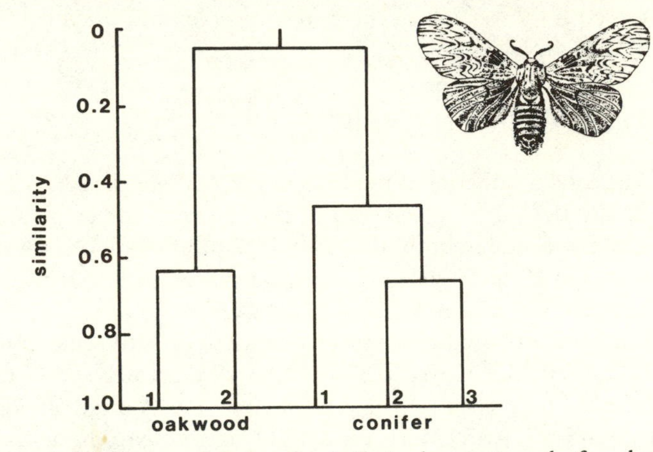

Figure 5.8 A dendrogram showing the similarity between moths found at three light-trap sites in Banagher conifer plantation and two light trap sites in Banagher oakwood. The cluster analysis was carried out using Jaccard's similarity coefficient and the group average method of agglomeration. The dendrogram shows much greater similarity (lower β diversity) within the two woodland habitats than between them.

with each other. Two of the most widely used methods in ecology are group average clustering and centroid clustering. An excellent discussion of cluster analysis is to be found in Pielou (1984).

Cluster analysis can be carried out using either presence and absence data or quantitative data. In many cases however (see for example Figure 5.9) the results are virtually identical. Since the interpretation of a cluster analysis depends on the visual inspection of the dendrogram the technique works best when performed with small data sets. A dendrogram of 30 sites or more is

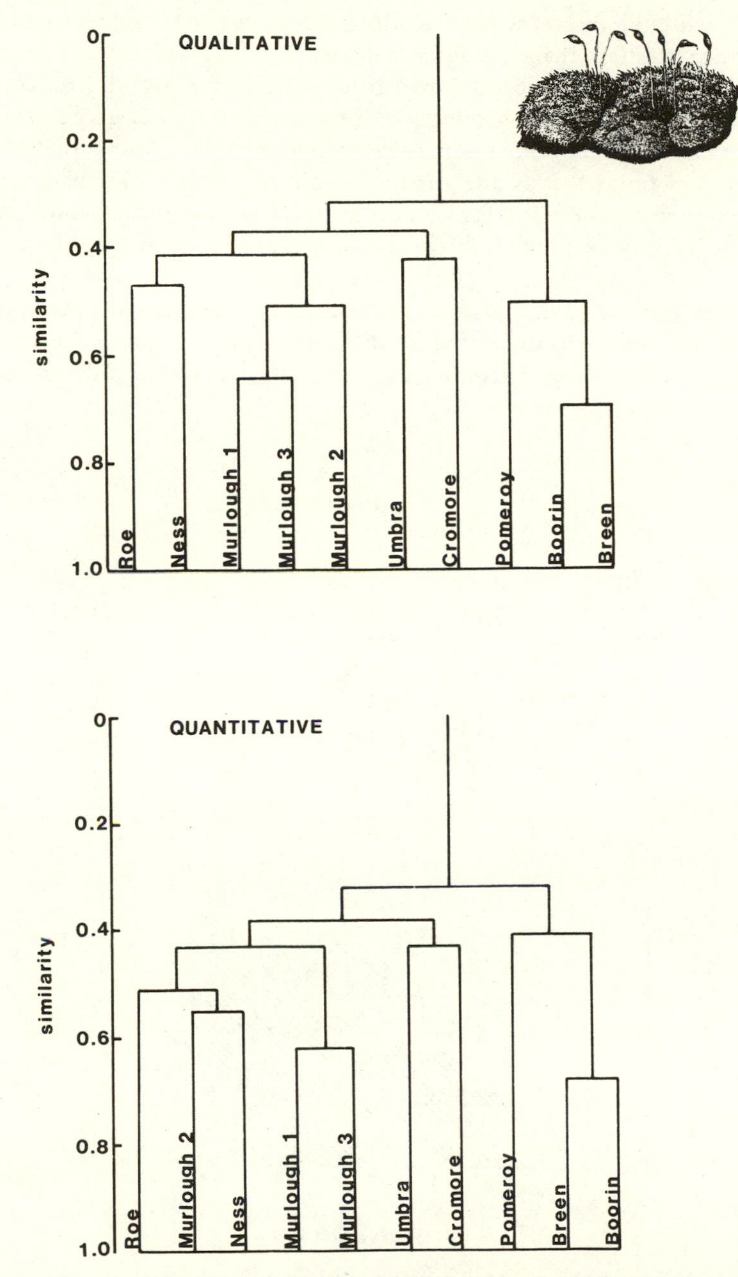

Figure 5.9 Dendrograms constructed using qualitative and quantitative data can often be quite similar. This graph shows the results of two cluster analyses (one using presence and absence data, the other abundance data) of ground vegetation in ten woodlands in Northern Ireland (Figure 6.6). The Jaccard and Sorensen (quantitative) indices were used.

often difficult to interpret while a dendrogram of over 100 sites is more likely to produce eyestrain than ecological insight!

Ordination techniques can be used to investigate the overall similarity of sites and to pick out major groupings. These methods do not give any direct measure of β diversity *per se* but may be used to infer the number of different communities present. It is also often possible to identify the characteristic species in each community. Two useful techniques are principal components analysis (Pielou, 1984; Jeffers, 1978) and indicator species analysis (Hill *et al.*, 1975).

One simple method of measuring β diversity is to examine the distribution of similarity coefficients calculated for different samples. Figure 5.10 contrasts β diversity in the Banagher conifer plantation and oakwood. β diversity was

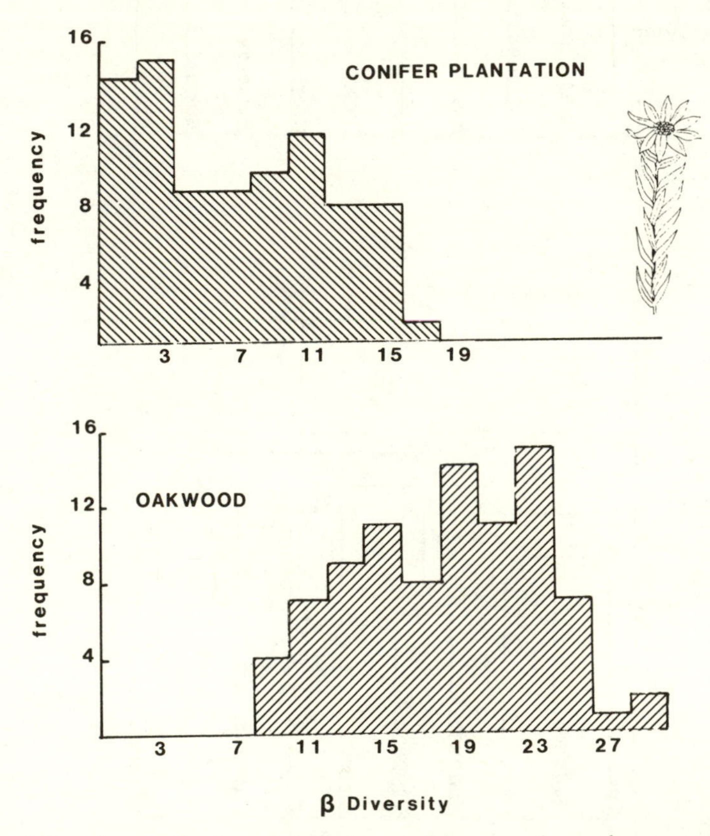

Figure 5.10 The distribution of similarity coefficients can be used as a measure of β diversity. The values of Jaccard's coefficient (weighed for species richness) is used to compare Banagher conifer plantation and Banagher oakwood. The latter is clearly more diverse.

calculated between successive quadrats along ten transects in each site using the equation:

$$\beta = (a + b) \times (1 - S) \tag{5.10}$$

where S = similarity calculated using the Jaccard index, a = the number of species in quadrat A, and b = the number of species in quadrat B.

The value of β increases as the number of species in the two quadrats increases and also as they become more dissimilar.

Summary

Measures of species diversity can be employed in other contexts. Two common applications involve investigations of habitat diversity and niche width (the diversity of resources which an organism or species utilizes). Like species diversity, these other forms of ecological diversity can be measured using either a simple richness index or a more complex index. Like species diversity measures these other measures of diversity are also subject to problems such as small sample bias. A system of habitat classification or resource classification must precede any study and it is important to take an organism's eye-view when this is being devised.

A second variety of ecological diversity concerns the degree of change in species composition between sites or communities or along gradients. This β or differentiation diversity can be described using similarity measures and the standard ecological techniques of classification and ordination. A number of special measures have been developed to measure species turnover along gradients.

6 The empirical value of diversity measures

The preceding chapters of this book have dealt primarily with the mechanical questions of calculating diversity indices, measuring abundance and determining sample size. Although the search for methods of measuring diversity is intellectually rewarding it is not a goal in itself. The true value of diversity measures will be determined by whether or not they are empirically useful.

There are two main areas in which diversity measures have potential application. These are in conservation, which is underpinned by the idea that species-rich communities are better than species-poor ones, and in environmental monitoring where the assumption that the adverse effects of pollution will be reflected in a reduction in diversity or by a change in the shape of the species abundance distribution is a central theme. In both cases diversity is used as an index of ecosystem wellbeing. As such it has great intuitive appeal. After all who could dispute the notion that greater diversity means higher ecological quality or deny that the use of a measure of diversity adds scientific rigor to a decision that might otherwise be made on subjective grounds alone! Yet the two areas differ in the way in which diversity measures are used. Environmental monitoring makes extensive use of diversity indices and species abundance distributions while conservation management concentrates almost exclusively on measures of species richness.

This chapter will examine the role of diversity measures in environmental monitoring and consider their potential application in conservation management. It will also point out instances where it may be misleading to base judgements on diversity indices without taking other ecological information into account.

Environmental assessment

The widely held assumption that diversity (good) will decrease with pollution (bad) has led to the use of diversity measures as environmental indicators. Nearly every index and model has been tried at one time or another and opinions differ widely as to which is the best buy. At one end of the spectrum are those ecologists who prefer to examine the full shape of the species abundance distribution while at the other are those who favour simple richness or dominance measures. There is general consensus however that enriched or polluted systems display a reduction in diversity (Rosenberg, 1976; Schafer, 1973; Wu, 1982). May (1981) noted that stable, equilibrium communities often follow a log normal pattern of species abundance. He further observed that when a mature community becomes polluted its species abundance distribution shifts backwards through succession to take up the shape of the less equitable log or geometric series. Classic data by Patrick (1973 and Figure 6.1) illustrate the point nicely by showing the effect of organic pollution on the diversity of a diatom community. The Park Grass experiment forms another excellent example. The Park Grass consists of a series of plots of permanent old pasture at Rothamsted, England, which have been subjected to various treatments for a century or more (Kempton, 1979; May, 1981). On one plot, which had been given a continuous heavy application of nitrogen, species richness decreased from 49 species in 1856 to three species in 1949, the

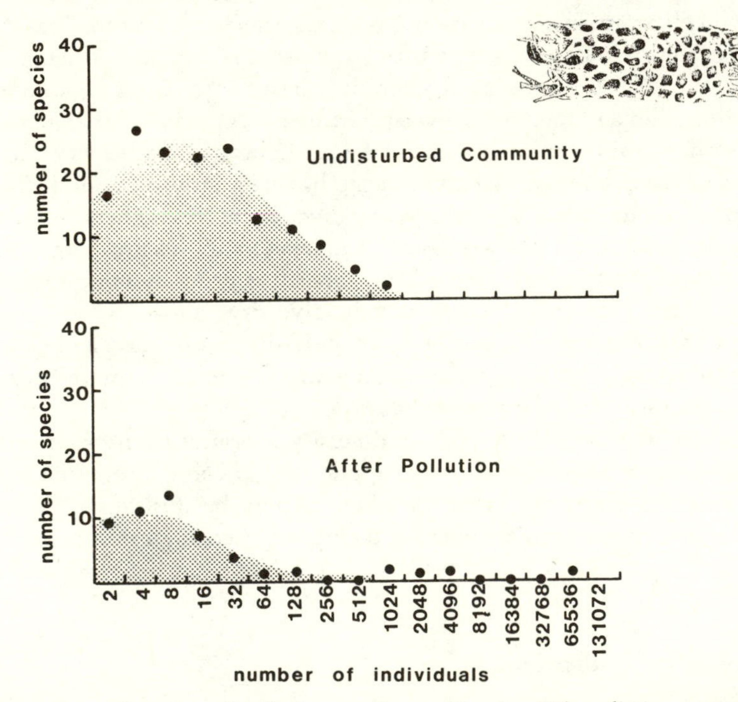

Figure 6.1 The effect of pollution on diatom diversity. When diatom communities are exposed to organic pollution the classic log normal distribution is replaced by one more reminiscent of the geometric or log series distribution found in immature or stressed communities. Abundancies are plotted in $\log_2$ and in all cases the upper bound of each abundance class (2, 4, 8, etc.) is shown. Redrawn from May (1981) after Patrick (1973).

percentage dominance of the commonest species rose from 14.5% to 99.7% and the species abundance distribution slipped back from log normal to geometric series.

Gray and Mirza (1979) and Ugland and Gray (1982) have also supported the idea that pollution-induced disturbance can be monitored by a departure from a log normal distribution to one where there is increased dominance. Shaw *et al.* (1983) and Lambshead and Platt (1985) however dispute the assertion that log normal distributions are universally present in equilibrium communities and universally absent from stressed ones. They note too that it can be difficult to choose between the log normal and other models when the distribution is truncated (see Chapter 4). Instead Shaw *et al.* (1983) plump for a Berger–Parker-style dominance index and show (Figure 6.2) that it can register the effect of organic effluent on the diversity of macrobenthos. Lambshead *et al.* (1983) also favour the use of dominance to rank communities under stress.

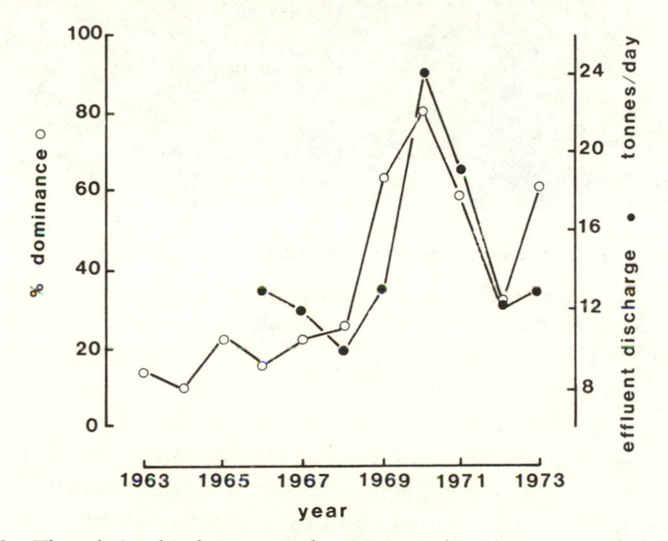

Figure 6.2 The relationship between a dominance index (the percentage abundance of the most abundant species) and effluent discharge (organic enrichment from a pulp mill) in Loch Eil on the west coast of Scotland. As the value of the dominance index increases the diversity of macrobenthos in the loch diminishes. Redrawn from Lambshead and Platt (1983) after Pearson (1975).

Tomascik and Sander (1987) were interested in the effects of eutrophication on reef-building corals in Barbados, West Indies. They found that eutrophication processes, in the form of nutrient enrichment, sedimentation, turbidity, toxicity and bacterial action, both directly and indirectly affected the community structure of the scleractinian coral assemblages. The best and most sensitive measures of these effects were provided by diversity indices. A range of indices were tested. The Shannon, Brillouin and Margalef indices produced largely equivalent findings. This accords with the discussion of the measures in Chapter 4. The Simpson index proved useful at detecting shifts in dominance. Shannon evenness measures produced the lowest degree of discrimination between the coral assemblages. Interestingly, Shannon evenness measures which used the amount of coral cover as the measure of abundance were poorer discriminators than those which took abundance as the number of colonies of coral. The relationship between the Brillouin index and eutrophication is illustrated in Figure 6.3.

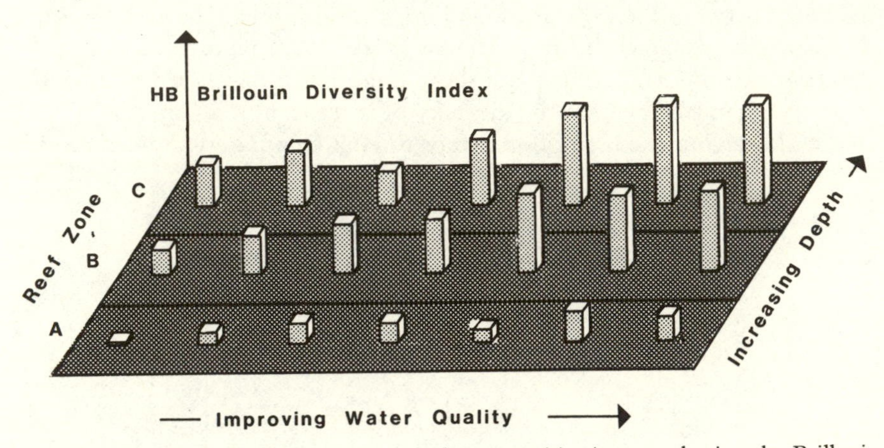

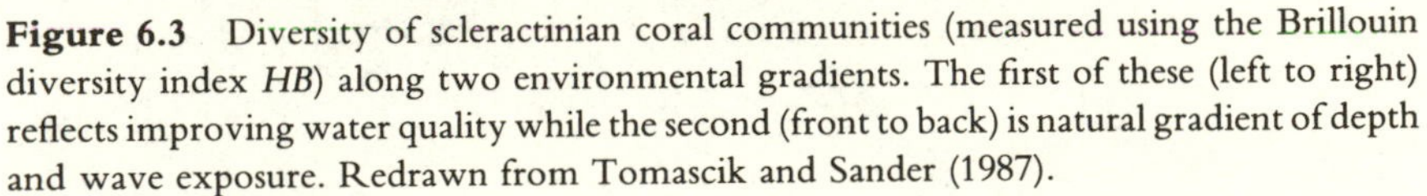

Figure 6.3 Diversity of scleractinian coral communities (measured using the Brillouin diversity index *HB*) along two environmental gradients. The first of these (left to right) reflects improving water quality while the second (front to back) is natural gradient of depth and wave exposure. Redrawn from Tomascik and Sander (1987).

Rosenzweig (1971) with his 'paradox of enrichment' and Tilman (1982) with his theory of resource competition have put forward ideas to explain why an increase in productivity should lead to a reduction in diversity.

A whole range of measures have been used in environmental assessment. Given the popularity of the Shannon index it is not surprising that it is widely adopted in pollution monitoring. Bechtel and Copeland (1970) showed that the diversity of fish in Galveston Bay, Texas, increased with increasing distance from Baytown, the site of considerable effluent discharge (Figure 6.4). Egloff and Brakel (1973) used the Shannon index to monitor the change in the diversity of benthic macroinvertebrates along an Ohio stream. Diversity dropped dramatically below a sewage outfall. This occurred irrespective of whether diversity was calculated at the level of the genus, order or class. Other water-quality parameters, for instance BOD (biological oxygen demand) and faecal coliform counts, paralleled the change in diversity.

The Shannon index was also employed by Wu (1982) who was interested in

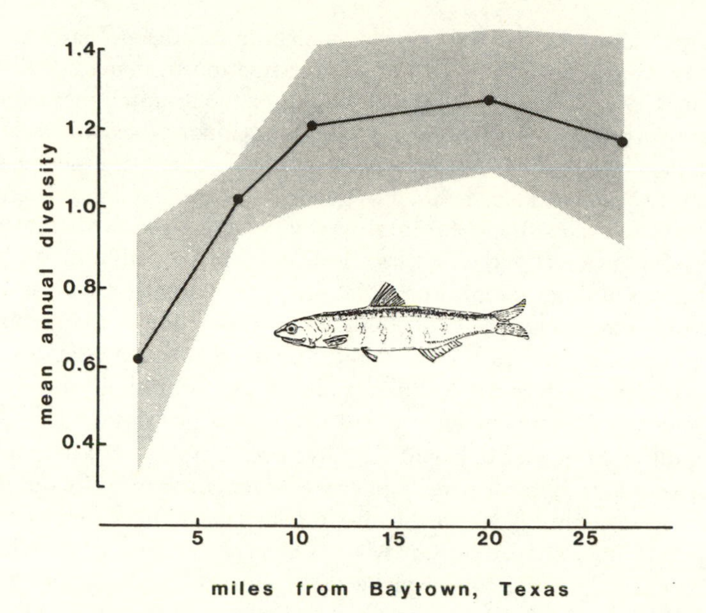

Figure 6.4 Fish diversity and pollution. This figure (redrawn from Bechtel and Copeland, 1970) shows that the diversity (measured using the Shannon index H': 95% confidence limits are also shown) of fish increases with distance from Baytown, Texas.

epibenthic communities in Tolo Harbour and Channel in Hong Kong. This is a subtropical environment subjected to a gradient of organic pollution. Wu found a clear increase in diversity with increasing distance from the pollution source.

Poiner and Kennedy (1984) used the Shannon index to measure the impact of dredging on the marine benthos of a large tropical sublittoral sandbank off Queensland, Australia. They recorded a significant decrease in diversity in the dredged areas. The surrounding non-dredged areas showed an increase in species richness, but not in diversity as measured by the Shannon index.

Mason (1977) was not altogether satisfied with the Shannon index when he used it to compare the diversity of macrobenthos in two shallow lakes in East Anglia, England. One of the lakes was eutrophic, the other unpolluted. Although Mason found that the Shannon index did discriminate between the two sites he obtained a more consistent difference using species richness alone.

Other researchers have found species richness a perfectly satisfactory measure of the effects of stress. Cairns (1969) recorded a large reduction in the number of species of protozoa in plastic troughs after temperature and pH shock while Homer (1976) noted major differences between the number of species of fish per 1000 individuals found in two adjacent Florida salt marshes, one of which received thermal pollution from a power station.

Simpson's index has also been employed in biomonitoring. Platt *et al.* (1984) preferred it to the Shannon index in an investigation of nematode diversity.

Taylor (1986) has used the log series index, α, to monitor the diversity of moths at many sites across Britain in relation to habitat type, latitude and land use. This data base will be used to forecast the effects of environmental change.

Indicator species can also be used to gauge environmental degradation and are particularly valuable when employed in conjunction with measures of diversity. Stoermer (1984) discusses the role of phytoplankton species and assemblages as biological indicators. Diatoms are potentially the most useful of the phytoplankton since they are abundant in most bodies of water and well preserved in sediments due to their silica 'skeleton'. An investigation of the turnover of diatom species in sediments can provide an insight into a range of environmental problems including acid rain (Battarbee *et al.*, 1985).

Many different groups of organisms have been employed in palaeoecology and palaeontology to reconstruct past environments and measure the effects of climatic and agricultural change. Fossilized pollen grains have for instance been used to build up a picture of East African Vegetation during the last ice age (Hamilton, 1982) while fossilized beetle remains demonstrate how insects responded to the dramatic changes of the Quaternary (Coope, 1978).

Platt *et al.* (1984) warn against the use of single species indicators. They note that the abundance of organisms can vary according to factors other than degree of pollution, even in a species noted for its sensitivity to pollution. They also observe that in many groups, including their own specialization of nematodes, there are no candidate indicator species.

Which approach is best?

The above studies clearly indicate that diversity measures have an important role in environmental assessment. They also confirm the conclusions drawn in Chapter 4. First, it is often useful to look at the change in the overall species abundance distribution. This observation is especially pertinent when polluted or enriched communities are under consideration. Next, simple species richness and dominance measures are invariably informative. Although seldom used, the Margalef index could be an important tool in this context. Thirdly, the Shannon index is fashionable and often useful, but, as a few of the examples above have indicated, it can be less informative than a simpler species richness measure. Fourthly, the log series index α is, like the Margalef index, only infrequently applied. Yet the extensive research into its properties (see Chapter 4) suggests that it could be a valuable measure in assessment work. Ideally α should replace the Shannon index as the preferred measure. Finally, indicator species are a useful adjunct to investigations of diversity. They can provide an additional clue into the way in which the community structure is changing.

Interpretation of results

Green and Vascotto (1978) and Green (1979) have suggested that diversity measures are an inappropriate way of measuring the effects of pollution. This conclusion is partly based on the observation that a number of studies have shown that diversity can be dependent on factors other than pollution (Bouchon, 1981; Loya, 1972). As with any ecological study it is important to distinguish causation and correlation. The observation that diversity increases as pollution decreases does not automatically prove that the one is a direct response to the other. Care is therefore needed when interpreting the results of studies similar to those described above.

It is also worth asking whether an increase in species diversity is actually an indication of increasing environmental quality. Initially, enrichment may cause an increase in diversity (Tilman, 1982) but this can be at the cost of a shift in the composition of the community. For instance an oligotrophic lake experiencing moderate inputs of phosphates and nitrates may acquire more species. But is this a sign that it is a better system? It is obvious that those involved in environmental assessment must be clear about what they mean by environmental quality.

Conservation and nature reserve management

There can be no doubt that diversity is a central concern of conservationists.

In *The Nature Conservation Review* Ratcliffe (1977) states that: 'diversity can be measured as an attribute and as such has neutral value; but because high diversity usually has more interest to biologists than low diversity the actual value measured can be used as a measure of quality in this respect'.

This application of diversity as an 'analogue' of conservation value (Rose, 1978; Yapp, 1979) is a common feature of ecological evaluation. Margules and Usher (1981) for instance examined nine published schemes concerned with the assessment of conservation potential and ecological value. In each case Margules and Usher listed the criteria used to judge the suitability of a habitat for conservation. Diversity emerged as the most widely used criterion; it appeared in eight out of the nine schemes. Rarity was also important. A follow up survey carried out by Margules (cited in Usher, 1986) extended the scope of this 'popularity poll' to 17 evaluation schemes (Table 6.1). Of all the 24 criteria listed, diversity was by far the most widely used.

Table 6.1 Popularity of criteria used in 17 conservation evaluation schemes. Diversity, the most frequently adopted criterion, appears in all but one scheme. From Usher (1986).

Criteria	*Frequency of use*
Diversity (of habitats and/or species)	16
Naturalness, rarity (of habitats and/or species)	13
Area	11
Threat of human interference	8
Amenity and educational value, representativeness	7
Scientific value	6
Recorded history	4
Population size, typicalness	3
Ecological fragility, position in ecological/geographical unit, potential value, uniqueness	2
Archaeological interest, availability, importance for migratory wildfowl, management factors, replaceability, silvicultural gene bank, successional stage, wildlife reservoir potential	1

What do conservationists mean by diversity?

Conservationists almost invariably view species diversity as species richness (see Norton, 1986). This is usually based on the rationale that species have the right to exist (Ehrenfeld, 1976) or that they have an actual or potential economic benefit to man (Frankel and Soulé, 1981; Everett, 1978; Helliwell, 1973, 1982). The preservation of genetic diversity is another frequent concern. Vida (1978) has stressed the importance of conserving polymorphisms and Harris *et al.* (1984) warn of the dangers of inbreeding in populations isolated in nature reserves.

Maximizing diversity

Considerable effort has been devoted to devising schemes that maximize the diversity of nature reserves. Much of this derives from the principles embodied in the Theory of Island Biogeography, proposed by MacArthur and Wilson (1967). (See also Gorman, 1979 and Williamson, 1981.) These guidelines for the selection and design of nature reserves (discussed by Diamond, 1975; Diamond and May, 1981; Game and Peterken, 1984; Harris, 1984; Higgs, 1981; Higgs and Usher, 1980; Janzen, 1983; Pickett and Thompson, 1978; Simberloff, 1986; Simberloff and Abele, 1982; Simberloff and Gotelli, 1984; Soulé and Wilcox, 1980; Terborgh, 1975 and Wilson and Willis, 1975) treat diversity purely as species richness.

Although assessment schemes, such as those reviewed by Margules and Usher (1981) and Usher (1986), and the systems of nature reserve design referred to above, consider diversity to be very important it would be most misleading to use diversity as the sole criterion or consider it independently from the type of habitat to be conserved (Margules, 1986). If, to give a rather obvious example, nature reserves were declared solely on the basis of species richness, important but species-poor habitats (such as salt marsh or upland woodland in Britain) might never be conserved. Figure 6.5 gives an example.

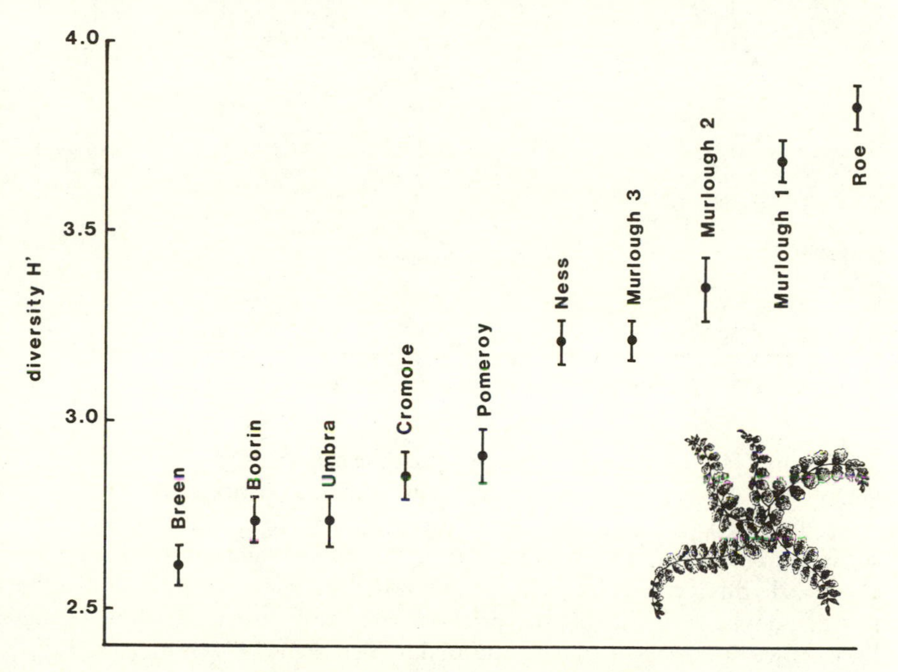

Figure 6.5 Assessing sites by diversity alone can be misleading. The diversity of ground vegetation in ten woodlands (see Figure 6.6) was estimated using the Shannon index H′ and used to rank the sites from the most diverse to the least diverse. The two nature reserves in the sample, Boorin and Breen woods, have the lowest diversity!

The variety and abundance of ground flora in ten small woodlands in Northern Ireland (Figure 6.6) and diversity was estimated using the Shannon index. The two woodlands in the sample which are nature reserves and which were chosen as such because their vegetation is most characteristic of the natural woodland of the area come bottom of the list. The other woodlands are more diverse due to a combination of factors including size, geology and degree of human disturbance. Diversity therefore is only of value when it is used to compare like habitats, and when the effects of area have been accounted

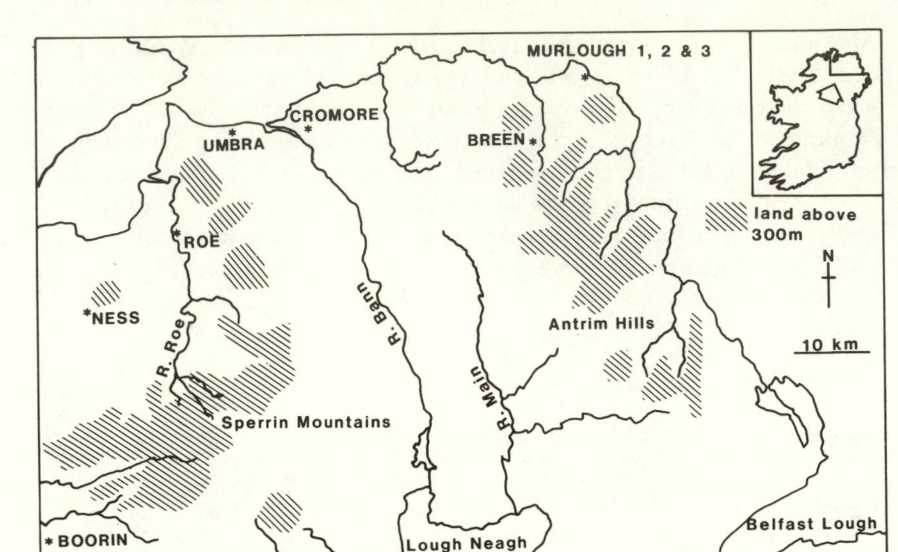

Figure 6.6 The locations of the ten woodlands sampled in Northern Ireland.

for (Figure 6.7). Ratcliffe (1986) stresses that the concept of diversity is only of value when it is applied to species characteristic of a particular ecosystem.

Let us pursue the woodland example a little further and assume that a conservation body decided to create a number of woodland nature reserves. In order to do this it could either choose the most species-rich sites irrespective of woodland type or first classify the different woodlands into stand groups [using for instance the excellent scheme outlined by Peterken (1974, 1981)] and then, all other things being equal, select the most diverse site or sites within each group as nature reserves. A series of species-rich nature reserves restricted to one or two woodland types (for example mixed woods on southern limestone: Figure 6.7) would be the result of the first approach. By contrast if the second approach was adopted the nature reserves would conserve a greater variety of woodland types, and because so many species have specialized habitat requirements, a greater overall species diversity. This point is expanded by Margules *et al.* (1982) who argue that genetic diversity is maximized not by using species richness as the sole criterion in site evaluation, but by making sure that 'whole suites of species' are conserved.

The practice of selecting representative examples of the whole range of natural habitats is now a well established part of conservation practice. Austin and Margules (1986) advocate the use of numerical classification as a method of subdividing the environment so that representative samples can be selected.

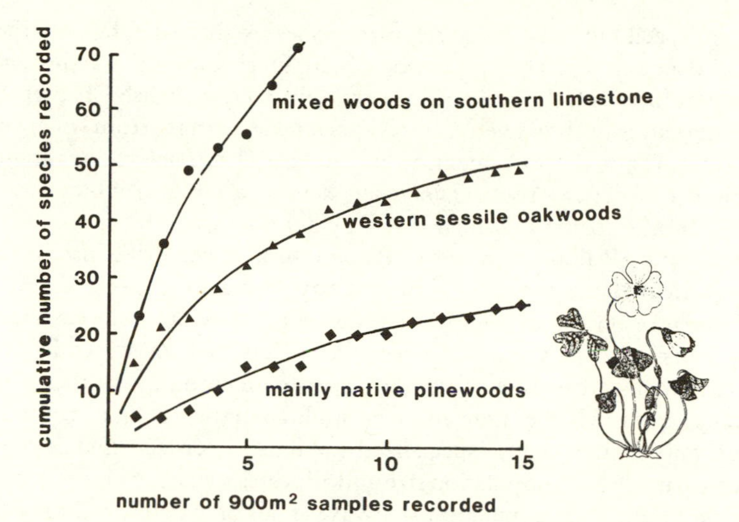

Figure 6.7 The relationship between the diversity of vascular plants (cumulative species richness) and the number of 900 m^2 samples in three types of woodland in Britain. Diversity increases with sample size and varies considerably between the woodland types in the comparison. Redrawn from Peterken (1981).

Austin and Margules go on to point out that in geographical areas which have not been intensively studied (for example Australia) representativeness may be a much more important criterion than either diversity or rarity. By taking a national perspective, and conserving a variety of habitats with different associated species, it is likely that the number of species conserved will in fact be maximized.

In assessing the diversity of a site for ecological evaluation all the considerations outlined in Chapter 3 must be borne in mind. For example assessment schemes are often limited to one or two groups of organisms and species lists may be incomplete (Spellerberg, 1981; Kirby *et al.*, 1986). An attempt to look at species abundances, as well as species richness, can be labour intensive, especially if there are many sites to be surveyed.

Rarity and conservation

Rarity followed a close second to diversity in Margules and Usher's (1981) survey of criteria for ecological assessment and is used in schemes evaluating a whole range of organisms [for example birds (Fuller and Langslow, 1986) and invertebrates (Disney, 1986)] in a variety of countries [including Scotland (Idle, 1986) and the Netherlands (van der Ploeg, 1986)]. Like so many other concepts associated with diversity, rarity has more than one meaning. In the context of species abundance models rare species are those that fall into the first few abundance classes. The proportion of rare species will decline through the sequence of models from geometric series to broken stick (see Chapter 2). To the conservationist however a rare species can range from one that is endangered and warrants a mention in the *Red Data Book* to a vagrant (for example the American robin *Turdus migratorius* in Britain) which strays far from its natural habitat in which it is very abundant.

Rabinowitz (Rabinowitz, 1981; Rabinowitz *et al.*, 1986) has devised a scheme which clarifies the concept of rarity by partitioning species distribution and abundance on three scales. First using geographic area she distinguishes species which occur over a large area from those which are endemic to a restricted area. Next she subdivides species according to their habitat specificity, that is whether they are cosmopolitan in their habitat requirements or exist only within a few specialized habitats. Then she makes the final dichotomy using local population size and allocates species to classes according to whether their local population is always small or whether it can be large. One cell represents common species. These are species which have a wide geographic range, large local population size and are found in a range of habitat types. The remaining categories describe seven different types of rarity. Rabinowitz and her co-workers asked a group of ecologists and systematists to assign the 177 species of native British flora described in detail in *The Biological Flora of the British Isles* (British Ecological Society, 1975, *et seq.*) to the eight cells in the $2 \times 2 \times 2$ table. The results for the 160 species for which there was no ambiguity are shown in Table 6.2. Species were not divided equally between the seven classes of rarity. One cell (small population, broad habitat requirements, narrow geographic range) contained no species at all. The majority of remaining rare species were placed in the restricted habitat category. This result shows that the conservationist preoccupation with the preservation of particular habitat types is justified since this strategy ensures that the largest number of rare species will be conserved. Rabinowitz, Cairns

Table 6.2 Rabinowitz's three-way classification of rarity: 160 species of plant described in the British Flora are allocated to the eight cells in the analysis. Common species are to be found in the cell with wide geographic distribution, large population size and broad habitat specificity. The remaining seven cells represent seven forms of rarity.

Geographic distribution	*Wide*		*Narrow*	
Habitat specificity	*Broad*	*Restricted*	*Broad*	*Restricted*
Local population size				
Somewhere large	58	71	6	14
Everywhere small	2	6	0	3

and Dillon conclude that their quantification of rarity will greatly facilitate the conservation of rare species. They also draw encouragement from the finding that their different judges were consistent in their classification of species on the basis of range, habitat and population size.

The role of diversity measures in conservation

A recent study has demonstrated that conservationists do indeed have something to gain from taking account of the relative abundances of species as well as their variety. Great concern has been voiced at the destruction of the tropical rain forests. These forests are uniquely species rich. But this diversity is more than just a preponderance of species. The real ecological puzzle lies in the fact that so many of the trees found in these tropical rain forests are rare. Hubbell and Foster (1986) looked at the pattern of species abundances in the forest on Barro Colorado Island, Panama, with particular emphasis on the commonness and rarity of species. They discovered that the species abundances of trees on the island were not log normally distributed. There were too many rare species for this to be the case. Hubbell and Foster made seven recommendations for tropical tree conservation based on their observations of the ecology of the rare species. These recommendations differ in a number of ways from those typically made in schemes which concentrate solely on species richness.

Why use diversity measures?

Some ecologists reject diversity indices and the use of species abundance distributions in favour of simple counts of numbers of species. In many cases species richness is an informative measure. Yet, as this book has I hope demonstrated, a considerable degree of ecological insight can be gleaned from more detailed investigations of the variety and abundances of species. In some cases a change in diversity, either by a shift in the species abundance distribution or an increase in dominance, will alert ecologists to detrimental processes such as pollution. In other instances greater information about the structure of different communities may be obtained from an examination of the relative abundances of species. Diversity measures are valuable, but are only a means to an end. That end is that ecologists should be able to ask the questions and formulate the hypotheses to help them understand, and sensibly manage, the natural world.

Summary

The major applications of diversity measurement are in nature conservation and environmental monitoring. In both cases diversity is held to be synonymous with ecological quality. Diversity measures are used extensively to gauge the adverse effects of pollution and environmental disturbance. Although there is considerable disagreement about which index or model is the most sensitive indicator of damage the general picture that emerges is that polluted or stressed environments experience a shift from a log normal pattern of species abundance, an increase in dominance and a decrease in species richness. Conservationists, who rate diversity most highly amongst their criteria for site assessment, concentrate almost exclusively on measures of species richness. There is however evidence that conservation strategies may be improved if information on species abundance patterns is taken into account. In all studies it is important to be clear whether an increase in diversity is the same as an increase in ecological quality.

References

Abbot, I. (1974) Numbers of plant, insect and land bird species on nineteen remote islands in the southern hemisphere. *Biol. J. Linn. Soc.*, **6**, 143–52.

Abele, L. G. (1974) Species diversity of decapod crustaceans. *Ecology*, **55**, 156–61.

Adams, J. E. and McCune, E. D. (1979) Application of the generalized jack-knife to Shannon's measure of information used as an index of diversity. In *Ecological Diversity in Theory and Practice* (eds J. F. Grassle, G. P. Patil, W. Smith and C. Taille), International Co-operative Publishing House, Fairland, MD, pp. 117–31.

Alatalo, R. and Alatalo, R. (1977) Components of diversity: multivariate analysis with interaction. *Ecology*, **58**, 900–6.

Allan, J. D. (1975) The distributional ecology and diversity of benthic insects in Cement Creek, Colorado. *Ecology*, **56**, 1040–53.

Anderson, M. C. (1964) Studies in woodland light climate. I. The photographic computation of light conditions. *J. Ecol.*, **542**, 27–41.

Anderson, M. C. (1971) Radiation and crop structure. In *Plant Photosynthetic Production Manual of Methods* (eds Z. Sestak, J. Catsky and P. G. Jarvis), Junk, The Hague, pp. 447–50.

Austin, M. P. and Margules, C. R. (1986) Assessing representativeness. In *Wildlife Conservation Evaluation* (ed. M. B. Usher), Chapman and Hall, London, pp. 46–67.

Battarbee, R. W., Flower, R. J., Stevenson, A. C. and Rippey, B. (1985) Lake acidification in Galloway: a palaeoecological test of competing hypotheses. *Nature*, **314**, 350–2.

Batten, L .A. (1976) Bird communities of some Kilarney woodlands. *Proc. Roy. Irish Acad.* **76**, 285–313.

Bechtel, T. J. and Copeland, B. J. (1970) Fish species diversity indices as indicators of pollution in Galveston Bay, Texas. *Contrib. Mar. Sci.*, **15**, 103–32.

Begon, M., Harper, J. L. and Townsend, C. R. (1986) *Ecology: Individuals, Populations and Communities*. Blackwell, Oxford.

Berger, W. H. and Parker, F. L. (1970) Diversity of planktonic Foraminifera in deep sea sediments. *Science*, **168**, 1345–7.

Blondel, J. and Cuvillier, R. (1977) Une methode simple et rapide pour decrire les habitats d'oiseaux: le stratiscope. *Oikos*, **29**, 326–31.

Boswell, M. T. and Patil, G. P. (1971) Chance mechanisms generating the logarithmic series distribution used in the analysis of numbers of species and individuals. In *Statistical Ecology* (eds G. P. Patil, E. C. Pielou and W. E. Waters), Pennsylvania State University Press, University Park, PA, pp. 99–130.

Bouchon, C. (1981) Quantitative study of the scleractinian coral communities on a fringing reef of Reunion Island (Indian Ocean). *Mar. Ecol. Prog. Ser.*, **4**, 273–88.

Bowman, K. O., Hutcheson, K., Odum, E. P. and Shenton, L. R. (1971) Comments on the distribution of indices of diversity. In *Statistical Ecology*, Vol. 3 (eds G. P. Patil, E. C. Pielou and W. E. Waters), Pennsylvania State University Press, University Park, PA, pp. 315–66.

Bray, J. R. and Curtis, C. T. (1957) An ordination of the upland forest communities of southern Wisconsin. *Ecol. Monogr.*, **27**, 325–49.

Brown, J. H. and Maurer, B. A. (1986) Body size, ecological dominance and Cope's rule. *Nature*, **324**, 248–50.

Brown, V. K. and Southwood, T. R. E. (1987) Secondary succession: patterns and strategies. In *Colonization, Succession and Strategies: 26th Symposium of the British Ecological Society*, Blackwell, Oxford, pp. 315–38.

Bulmer, M. G. (1974) On fitting the Poisson lognormal distribution to species abundance data. *Biometrics*, **30**, 101–10.

Bunce, R. G. H. and Shaw, M. W. (1973) A standardized procedure for ecological survey. *J. Environ. Manag.*, **1**, 239–85.

Burger, J. (1972) The use of a fish-eye lens to study nest placement in Franklin's gulls. *Ecology*, **53**, 362–4.

Cairns, J. (1969) Rate of species diversity restoration following stress in freshwater protozoan communities. *Univ. Kansas Sci. Bull.*, **48**, 209–24.

Chalmers, N. and Parker, P. (1986) *The OU Project Guide*. The Field Studies Council, Taunton.

Clements, F. E. (1916) Plant succession: an analysis of the development of vegetation. *Carneg. Instit. Wash. Publ.*, **242**, 1–512.

Clifford, H. T. and Stephenson, W. (1975) *An Introduction to Numerical Classification*, Academic Press, London.

Cody, M. L. (1975) Towards a theory of continental species diversity bird distributions over Mediterranean habitat gradients. In *Ecology and Evolution of Communities* (eds M. L. Cody and J. M. Diamond), Harvard University Press, Cambridge, Mass., pp. 214–57.

Cohen, A. C., Jr. (1961) Tables for maximum likelihood estimates: singly truncated and singly censored samples. *Technometrics*, **3**, 535–41.

Cohen, J. E. (1968) Alternative derivations of a species abundance relation. *Amer. Nat.*, **102**, 165–72.

Colwell, R. K. and Futuyama, D. J. (1971) On the measurement of niche breadth and overlap. *Ecology*, **52**, 567–76.

Connor, E. F. and Simberloff, D. S. (1978) Species number and compositional similarity of the Galapagos flora and avifauna. *Ecol. Monogr.*, **48**, 219–48.

Coope, G. R. (1978) Constancy of insect species versus inconstancy of quarternary environments. In *Diversity of Insect Faunas: 9th Symposium of the Royal Entomological Society* (eds L. A. Mound and N. Warloff), Blackwell, Oxford, pp. 176–87.

Currie, D. J. and Paquin, V. (1987) Large-scale biogeographical patterns of species richness of trees. *Nature*, **329**, 326–7.

De Caprariis, P. and Lindemann, R. (1978) Species richness in patchy environments. *Math. Geol.*, **10**, 73–90.

Diamond, J. M. (1975) The island dilemma: lessons of modern biogeographic studies for the design of natural reserves. *Biol. Conserv.*, **7**, 129–46.

Diamond, J. M. and May, R. M. (1981) Island biogeography and the design of nature reserves. In *Theoretical Ecology: Principles and Applications* (ed. R. M. May), Blackwell, Oxford, pp. 163–86.

Disney, R. H. L. (1986) Assessments using invertebrates: posing the problem. In *Wildlife Conservation Evaluation* (ed. M. B. Usher), Chapman and Hall, London, pp. 272–93.

Egloff, D. A. and Brakel, W. H. (1973) Stream pollution and a simplified diversity index. *J. Water Pollut. Contr. Fed.*, **45**, 2269–75.

Ehrenfeld, D. W. (1976) The conservation of non-resources. *Amer. Sci.*, **64**, 648–56.

Elliot, J. M. (1981) *Some Methods for the Statistical Analysis of Samples of Benthic Invertebrates*, Scientific Publications of the Freshwater Biological Association 25, 148 pp.

Elton, C. S. (1958) *The Ecology of Invasions by Animals and Plants*, Methuen, London.

Elton, C. S. (1966) *The Pattern of Animal Communities*, Methuen, London.

Elton, C. S. and Miller, R. S. (1954) The ecological survey of animal communities: with a practical system of classifying habitats by structural characteristics. *J. Ecol.*, **42**, 460–96.

Engen, S. (1978) *Stocastic Abundance Models with Emphasis on Biological Communities and Species Diversity*, Chapman and Hall, London.

Evans, G. C., Freeman, P. and Rackham, D. (1975) Developments in hemispherical photography. In *Light as an Ecological Factor II* (eds G. C. Evans, R. Bainbridge and O. Rackham), Blackwell, Oxford, pp. 549–56.

Everett, R. D. (1978) The wildlife preference shown by countryside visitors. *Biol. Conserv.*, **14**, 75–84.

Feinsinger, P., Spears, E. E. and Poole, R. W. (1981) A simple measure of niche breadth. *Ecology*, **62**, 27–32.

Fisher, R. A., Corbet, A. S. and Williams, C. B. (1943) The relation between the number of species and the number of individuals in a random sample of an animal population. *J. Anim. Ecol.*, **12**, 42–58.

Frankel, O. H. and Soulé, M. E. (1981) *Conservation and Evolution*, Cambridge University Press, Cambridge.

Frontier, S. (1985) Diversity and structure in aquatic ecosystems. In *Oceanography and Marine Biology, An Annual Review* (ed. M. Barnes), Aberdeen University Press, Aberdeen, pp. 253–312.

Fuller, R. J. and Langslow, D. R. (1986) Ornithological evaluation for wildlife conservation. In *Wildlife Conservation Evaluation* (ed. M. B. Usher), Chapman and Hall, London, pp. 247–69.

Game, M. and Peterken, G. F. (1984) Nature reserve selection strategies in the woodlands of central Lincolnshire, England. *Biol. Conserv.*, **29**, 157–81.

Gaudreault, A., Miller, T., Montgomery, W. L. and FitzGerald, G. J. (1986) Interspecific interactions and diet of sympatric juvenile brook charr *Salvelinus fontinalis*, and adult ninespine sticklebacks *Pungitius pungitius*. *J. Fish Biol.*, **28**, 133–40.

Giller, P. S. (1984) *Community Structure and the Niche*, Chapman and Hall, London.

Glime, J. M. and Clemons, R. M. (1972) Species diversity of stream insects on *Fontanalis* spp. compared to diversity on artificial substrates. *Ecology*, **53**, 458–64.

Goodman, D. (1975) The theory of diversity–stability relations in ecology. *Q. Rev. Biol.*, **50**, 237–66.

Gorman, M. L. (1979) *Island Ecology*, Chapman and Hall, London.

Gorman, O. T. and Karr, J. R. (1978) Habitat structure and stream fish communities. *Ecology*, **59**, 507–15.

Grassle, J. F., Patil, G. P., Smith, W. and Taille, C. (1979) *Ecological Diversity in Theory and Practice*, International Co-operative Publishing House, Fairland, MD.

Gray, J. S. (1988) Species abundance patterns. In *Organisation of Communities: Past and Present* (eds), Blackwell, Oxford.

Gray, J. S. and Mirza, F. B. (1979) A possible method for detecting pollution-induced disturbance on marine benthic communities. *Mar. Pollut. Bull.*, **10**, 142–6.

Green, R. H. (1979) *Sampling Design and Statistical Methods for Environmental Biologists*, Wiley, New York.

Green, R. H. and Vascotto, G. L. (1978) A method for the analysis of environmental factors controlling patterns of species composition in aquatic communities. *Water Res.*, **12**, 583–90.

Greig-Smith, P. (1983) *Quantitative Plant Ecology*, Blackwell, Oxford.

Hamilton, A. C. (1982) *Environmental History of East Africa: A Study of the Quaternary*, Academic Press, London.

Harberd, D. J. (1967) Observation on natural clones of *Holcus mollis*, *New Phytol.*, **66**, 401–8.

Harman, W. (1972) Benthic substrates: their effect on freshwater mollusca. *Ecology*, **53**, 271–7.

Harper, J. L. (1977) *Population Biology of Plants*, Academic Press, London.

Harper, J. L. (1981) The concept of population in modular organisms. In *Theoretical Ecology: Principles and Applications*, Blackwell, Oxford, pp. 53–77.

Harris, H. J., Milligan, M. S. and Fewless, G. A. (1983) Diversity: quantification and ecological evaluation in freshwater marshes. *Biol. Conserv.*, **27**, 99–110.

Harris, L. D., McGlothlen, M. E. and Manlove, M. N. (1984) Genetic resources and biotic diversity. In *The Fragmented Forest* (ed. L. D. Harris). University of Chicago Press, Chicago, pp. 93–107.

Harris, L. D. (1984) *The Fragmented Forest*, University of Chicago Press, Chicago.

Harvey, P. H. and Godfray, H. C. J. (1987) How species divide resources. *Amer. Nat.*, **129**, 318–20.

Harvey, P. H. and Lawton, J. H. (1986) Patterns in three dimensions. *Nature*, **324**, 212.

Hassell, M. P. and May, R. M. (1985) From individual behaviour to population dynamics. In *Behavioural Ecology* (eds R. Smith and R. Sibly), Blackwell, Oxford, pp. 3–32.

Helliwell, D. R. (1973) Priorities and values in nature conservation. *J. Environm. Manag.*, **1**, 85–127.

Helliwell, D. R. (1982) Assessment of conservation values of large and small organisms. *J. Environm. Manag.*, **15**, 273–7.

Heltshe, J. F. and Bitz, D. W. (1979) Comparing diversity measures in sampled communities. In *Ecological Diversity in Theory and Practice* (eds J. F. Grassle, G. P. Patil, W. Smith and C. Taille), International Co-operative Publishing House, Fairland, MD, pp. 133–44.

Hengeveld, R. (1979) On the use of abundance and species-abundance curves. In *Statistical Distributions in Ecological Work* (eds J. K. Ord, G. P. Patil and C. Taille), International Co-operative Publishing House, Fairland, MD, pp. 275–88.

Higgs, A. J. (1981) Island biogeography theory and nature reserve design. *J. Biogeogr.*, **8**, 117–24.

Higgs, A. J. and Usher, M. B. (1980) Should nature reserves be large or small? *Nature*, **285**, 568–9.

Hill, M. O. (1973) Diversity and evenness: a unifying notation and its consequences. *Ecology*, **54**, 427–31.

Hill, M. O., Bunce, R. G. H. and Shaw, M. W. (1975) Indicator species analysis: a divisive polythetic method of classification, and its application to a survey of native pinewoods in Scotland. *J. Ecol.*, **63**, 597–613.

Hill, R. (1924) A lens for whole sky pictures. *Q. J. R. Meteorol. Soc.*, **50**, 227–35.

Holloway, J. D. (1977) *The Lepidoptera of Norfolk Island: Their Biogeography and Ecology*, Junk, The Hague.

Homer, M. (1976) Seasonal abundance, biomass, diversity and trophic structure of fish in a salt-marsh tidal creek affected by a coastal power plant. In *Thermal Ecology II* (eds G. W. Esch and R. W. McFarlane), US Energy Research and Development Administration, Washington, DC, pp. 259–67.

Hopkins, B. (1957) The concept of minimal area. *J. Ecol.*, **45**, 441–9.

Hubbell, S. P. and Foster, R. B. (1986) Commonness and rarity in a neotropical forest: implications for tropical tree conservation. In *Conservation Biology: The Science of Scarcity and Diversity* (ed. M. J. Soulé), Sinauer, Massachusetts, pp. 205–31.

Hughes, R. G. (1984) A model of the structure and dynamics of benthic marine invertebrate communities. *Mar. Ecol. Prog. Ser.*, **15**, 1–11.

Hughes, R. G. (1986) Theories and models of species abundance. *Amer. Nat.*, **128**, 879–99.

Hurlbert, S. H. (1971) The non-concept of species diversity: a critique and alternative parameters. *Ecology*, **52**, 577–86.

Hurlbert, S. H. (1978) The measurement of niche overlap and some derivatives. *Ecology*, **59**, 67–77.

Hutcheson, K. (1970) A test for comparing diversities based on the Shannon formula. *J. Theor. Biol.*, **29**, 151–4.

Idle, E. T. (1986) Evaluation at a local scale: a region in Scotland. In *Wildlife Conservation Evaluation* (ed. M. B. Usher), Chapman and Hall, London, pp. 182–98.

Janson, S. and Vegelius, J. (1981) Measures of ecological association. *Oecologia*, **49**, 371–6.

Janzen, D. H. (1983) No park is an island: increase in interference from outside as park size decreases. *Oikos*, **41**, 401–10.

Jeffers, J. N. R. (1978) *An Introduction to Systems Analysis with Ecological Applications*, Arnold, London.

Karr, J. R. and Roth, R. R. (1971) Vegetational structure and avian diversity in several New World areas. *Amer. Nat.*, **105**, 423–5.

Kempton, R. A. (1979) Structure of species abundance and measurement of diversity. *Biometrics*, **35**, 307–22.

Kempton, R. A. and Taylor, L. R. (1974) Log-series and log-normal parameters as diversity determinants for the *Lepidoptera*. *J. Anim. Ecol.*, **43**, 381–99.

Kempton, R. A. and Taylor, L. R. (1976) Models and statistics for species diversity. *Nature*, **262**, 818–20.

Kempton, R. A. and Taylor, L. R. (1978) The Q-statistic and the diversity of floras. *Nature*, **275**, 252–3.

Kempton, R. A. and Wedderburn, R. W. M. (1978) A comparison of three measures of species diversity. *Biometrics*, **34**, 25–37.

Kershaw, K. A. and Looney, J. H. H. (1985) *Quantitative and Dynamic Plant Ecology*, Arnold, London.

King, C. E. (1964) Relative abundance of species and MacArthur's model. *Ecology*, **45**, 716–27.

Kirby, K. (1986) Forest and woodland conservation. In *Wildlife Conservation Evaluation* (ed. M. B. Usher), Chapman and Hall, London, pp. 202–21.

Kirby, K. J., Bines, T., Burn, A., Mackintosh, J., Pitkin, P. and Smith, I. (1986) Seasonal and observer differences in vascular plant records from British woodlands. *J. Ecol.*, **74**, 123–32.

Kotrschal, K. and Thomson, D. A. (1986) Feeding patterns in eastern tropical Pacific blennioid fishes (Teleostei: Tripterygiidae, Labrisomidae, Chaenopsidae, Blenniidae). *Oecologia*, **70**, 367–78.

Krebs, C. J. (1985) *Ecology: The Experimental Analysis of Distribution and Abundance*. Harper and Row, New York.

Krebs, J. R. and Davies, N. B. (1981) *An Introduction to Behavioural Ecology*, Blackwell, Oxford.

Krebs, J. R. and Davies, N. B. (1984) *Behavioural Ecology: An Evolutionary Approach*, Blackwell, Oxford.

Lambshead, J. and Platt, H. M. (1985) Structural patterns of marine benthic assemblages and their relationships with empirical statistical models. In *Proceedings of the 19th European Marine Biology Symposium, Plymouth, 1984* (ed. P. E. Gibbs), Cambridge University Press, Cambridge, pp. 371–80.

Lambshead, P. J. D., Platt, H. M. and Shaw, K. M. (1983) Detection of differences among assemblages of benthic species based on an assessment of dominance and diversity. *J. Nat. Hist., London*, **17**, 859–74.

Lawton, J. H. (1976) The structure of arthropod communities on bracken. *Bot. J. Linn. Soc.*, **73**, 187–216.

Lawton, J. H. (1978) Host-plant influences on insect diversity: the effects of space and time. In *Diversity of Insect Faunas* (eds L. A. Mound and N. Waloff), Blackwell, Oxford, pp. 105–215.

Lawton, J. H. (1984) Non-competitive populations, non-convergent communities and vacant niches: the herbivores of bracken. In *Ecological Communities: Conceptual Issues and the Evidence* (eds D. R. Strong, D. Simberloff, L. G. Abele and A. B. Thistle), Princeton University Press, Princeton, pp. 67–100.

Laxton, R. R. (1978) The measure of diversity. *J. Theor. Biol.*, **70**, 51–67.

Lewis, T. and Taylor, L. R. (1967) *An Introduction to Experimental Ecology*, Academic Press, London.

Lloyd, M. and Ghelardi, R. J. (1964) A table for calculating the 'equitability' component of species diversity. *J. Anim. Ecol.*, **33**, 217–55.

Lloyd, M., Inger, R. F. and King, F. W. (1968) On the diversity of reptile and amphibian species in a Bornean rain forest. *Amer. Nat.*, **102**, 497–515.

Loya, Y. (1972) Community structure and species diversity of hermatypic corals at Eilat, Red Sea. *Mar. Biol.*, **13**, 100–23.

MacArthur, R. H. (1957) On the relative abundance of bird species. *Proc. Nat. Acad. Sci., USA*, **43**, 293–5.

MacArthur, R. H. (1960) On the relative abundance of species. *Amer. Nat.*, **94**, 25–36.

MacArthur, R. H. (1965) Patterns of species diversity. *Biol. Rev*, **40**, 510–33.

MacArthur, R. H. and Horn, H. S. (1969) Foliage profile by vertical measurements. *Ecology*, **50**, 802–4.

MacArthur, R. H. and MacArthur, J. W. (1961) On bird species diversity. *Ecology*, **42**, 594–8.

MacArthur, R. H., Recher, H. F. and Cody, M. L. (1966) On the relation between habitat selection and species diversity. *Amer. Nat.*, **100**, 319–27.

MacArthur, R. H. and Wilson, E. O. (1967) *The Theory of Island Biogeography*, Princeton University Press, Princeton.

Magurran, A. E. (1981) *Biological Diversity and Woodland Management,* Unpublished D.Phil. thesis, New University of Ulster.

Magurran, A. E. (1985) The diversity of Macrolepidoptera in two contrasting woodland habitats at Banagher, Northern Ireland. *Proc. R. Ir. Acad.,* **85B**, 121–32.

Mandelbrot, B. B. (1977) *Fractals, Fun, Chance and Dimension.* W. H. Freeman, San Francisco.

Mandelbrot, B. B. (1982) *The Fractal Geometry of Nature,* Freeman, San Francisco.

Margalef, R. (1972) Homage to Evelyn Hutchinson, or why is there an upper limit to diversity. *Trans. Connect. Acad. Arts Sci.,* **44**, 211–35.

Margules, C. R. (1986) Conservation evaluation in practice. In *Wildlife Conservation Evaluation* (ed. M. B. Usher), Chapman and Hall, London, pp. 298–314.

Margules, C. R., Higgs, A. J. and Rafe, R. W. (1982) Modern biogeographic theory: are there any lessons for nature reserve design. *Biol. Conserv.,* **24**, 115–28.

Margules, C. and Usher, M. B. (1981) Criteria used in assessing wildlife conservation potential: a review. *Biol. Conserv.,* **21**, 79–109.

Mason, C. F. (1977) The performance of an index of diversity in describing the zoobenthos of two lakes. *J. Appl. Ecol.,* **14**, 363–7.

May, R. M. (1973) *Stability and Complexity in Model Ecosystems,* Princeton University Press, Princeton.

May, R. M. (1974) General introduction. In *Ecological Stability* (eds M. B. Usher and M. H. Williamson), Chapman and Hall, London, pp. 1–14.

May, R. M. (1975) Patterns of species abundance and diversity. In *Ecology and Evolution of Communities* (eds M. L. Cody and J. M. Diamond), Harvard University Press, Cambridge, MA, pp. 81–120.

May, R. M. (1981) Patterns in multi-species communities. In *Theoretical Ecology: Principles and Applications* (ed. R. M. May), Blackwell, Oxford, pp. 197–227.

May, R. M. (1984) An overview: real and apparent patterns in community structure. In *Ecological Communities: Conceptual Issues and the Evidence* (eds D. R. Strong, D. Simberloff, L. G. Abele and A. B. Thistle), Princeton University Press, Princeton, pp. 3–18.

May, R. M. (1986) The search for patterns in the balance of nature: advances and retreats. *Ecology,* **67**, 1115–26.

McIntosh, R. P. (1967) An index of diversity and the relation of certain concepts to diversity. *Ecology,* **48**, 392–404.

Minshall, G. W., Petersen, R. C. and Nimz, C. F. (1985) Species richness in streams of different size from the same drainage basin. *Amer. Nat.,* **125**, 16–38.

Moore, P. D. and Chapman, S. B. (1985) *Methods in Plant Ecology,* Blackwell, Oxford.

Morris, M. G. and Lakhani, K. H. (1979) Responses of grassland invertebrates to management by cutting. I. Species diversity of Hemiptera. *J. Appl. Ecol.,* **16**, 77–98.

Moss, D. (1978) Diversity of woodland song-bird populations. *J. Anim. Ecol.,* **47**, 521–7.

Moss, D. (1979) Even-aged plantations as a habitat for birds. In *The Ecology of Even-Aged Forest Plantations* (eds E. D. Ford, D. C. Malcolm and J. Atterson), Institute of Terrestrial Ecology, Cambridge, pp. 413–28.

Motomura, I. (1932) A statistical treatment of associations [in Japanese and cited in May, 1975]. *Jpn. J. Zool.,* **44**, 379–83.

Mueller-Dombois, D. and Ellenberg, H. (1974) *Aims and Methods of Vegetation Ecology,* Wiley, New York.

Norton, B. G. (1986) *The Preservation of Species*, Princeton University Press, Princeton.

Odum, E. P. (1968) Energy flow in ecosystems: a historical review. *Amer. Zool.*, **8**, 11–18.

Patrick, R. (1973) Use of algae, especially diatoms, in the assessment of water quality. *American Society for Testing and Materials, Special Technical Publication 528*, 76–95.

Pearsall, S. H., Durham, D. and Eagar, D. C. (1986) Evaluation methods in the United States. In *Wildlife Conservation Evaluation* (ed. M. B. Usher), Chapman and Hall, London, pp. 111–34.

Pearson, T. H. (1975) The benthic ecology of Loch Linnhe and Loch Eil, a sea loch system on the west coast of Scotland. IV. Changes in the benthic fauna attributable to organic enrichment. *J. Exp. Mar. Biol. Ecol.*, **20**, 1–41.

Peet, R. K. (1974) The measurement of species diversity. *Ann. Rev. Ecol. System.*, **5**, 285–307.

Peterken, G. F. (1974) A method for assessing woodland flora for conservation using indicator species. *Biol. Conserv.*, **6**, 239–45.

Peterken, G. F. (1981) *Woodland Conservation and Management*, Chapman and Hall, London.

Phillips, D. J. H. (1980) *Quantitative Aquatic Biological Indicators*, Applied Science Publishers, Essex.

Pianka, E. R. (1966) Latitudinal gradients in species diversity: a review of concepts. *J. Amer. Nat.* **100**, 33–46.

Pianka, E. R. (1983) *Evolutionary Ecology*, Harper and Row, New York.

Pickett, S. T. A. and Thompson, J. N. (1978) Patch dynamics and the design of nature reserves. *Biol. Conserv.*, **13**, 27–37.

Pielou, E. C. (1966) Species diversity and pattern diversity in the study of ecological succession. *J. Theor. Biol.*, **10**, 370–83.

Pielou, E. C. (1969) *An Introduction to Mathematical Ecology*, Wiley, New York.

Pielou, E. C. (1975) *Ecological Diversity*, Wiley, New York.

Pielou, E. C. (1984) *The Interpretation of Ecological Data*, Wiley, New York.

Pielou, E. C. and Arnason, A. N. (1965) Correction to one of MacArthur's species-abundance formulas. *Science*, **151**, 592.

Pimm, S. L. (1982) *Food Webs*, Chapman and Hall, London.

Pimm, S. L. (1984) The complexity and stability of ecosystems. *Nature*, **307**, 321–6.

Platt, H. M., Shaw, K. M. and Lambshead, P. J. D. (1984) Nematode species abundance patterns and their use in the detection of environmental perturbations. *Hydrobiologia*, **118**, 59–66.

Poiner, I. R. and Kennedy, R. (1984) Complex patterns of change in the macrobenthos of a large sandbank following dredging. *Mar. Biol.*, **78**, 335–52.

Poole, R. W. (1974) *An Introduction to Quantitative Ecology*, McGraw-Hill Kogakusha, Tokyo.

Pope, D. J. and Lloyd, P. S. (1975) Hemispherical photography, topography and plant distribution. In *Light as an Ecological Factor II* (eds G. C. Evans, R. Bainbridge and O. Rackham), Blackwell, Oxford, pp. 385–408.

Preston, F. W. (1948) The commonness, and rarity, of species. *Ecology*, **29**, 254–83.

Preston, F. W. (1962) The canonical distribution of commonness and rarity. *Ecology*, **43**, 185–215 and 410–32.

Quenouille, M. H. (1956) Notes on bias in estimation. *Biometrika*, **43**, 353–60.

Rabinowitz, D. (1981) Seven forms of rarity. In *The Biological Aspects of Rare Plant Conservation* (ed. H. Synge), John Wiley, Chichester, pp. 205–17.

Rabinowitz, D., Cairns, S. and Dillon, T. (1986) Seven forms of rarity and their frequency in the flora of the British Isles. In *Conservation Biology: The Science of Scarcity and Diversity* (ed. M. J. Soulé), Sinauer, Sunderland, MA, pp. 182–204.

Ratcliffe, D. A. (1977) *A Nature Conservation Review*, Vols 1 and 2, Cambridge University Press, Cambridge.

Ratcliffe, D. A. (1986) Selection of important areas for wildlife conservation in Great Britain: the Nature Conservancy Council's approach. In *Wildlife Conservation Evaluation* (ed. M. B. Usher), Chapman and Hall, London, pp. 136–59.

Recher, H. F. (1969) Bird species diversity and habitat diversity in Australia and North America. *Amer. Nat.*, **103**, 75–80.

Reichelt, R. E. and Bradbury, R. H. (1984) Spatial patterns in coral reef benthos: multiscale analysis of sites from three oceans. *Mar. Ecol. Progr. Ser.*, **17**, 251–7.

Roberts, C. M. and Ormond, R. F. G. (1987) Habitat complexity and coral reef fish diversity and abundance on Red Sea fringing reefs. *Mar. Ecol. Progr. Ser.*, **41**, 1–8.

Rose, C. I. (1978) A note on diversity and conservation. *Bull. Brit. Ecol. Soc.*, **9**(4), 5–6.

Rosenberg, R. (1976) Benthic faunal dynamics during succession following pollution abatement in a Swedish estuary. *Oikos*, **27**, 414–27.

Rosenzweig, M. (1971) Paradox of enrichment: destabilization of exploitation ecosystems in ecological time. *Science*, **171**, 385–7.

Routledge, R. D. (1977) On Whittaker's components of diversity. *Ecology*, **58**, 1120–7.

Routledge, R. D. (1979) Diversity indices: which ones are admissible. *J. Theor. Biol.*, **76**, 503–15.

Russell, G. and Fielding, A. J. (1981) Individuals, populations and communities. In *The Biology of Seaweeds* (eds C. S. Lobban and M. J. Wynne), Blackwell, Oxford, pp. 393–420.

Sanders, H. L. (1968) Marine benthic diversity: a comparative study. *Amer. Nat.*, **102**, 243–82.

Schafer, C. T. (1973) Distribution of foraminifera near pollution sources in Chaleur Bay. *Water Air Soil Pollut.*, **2**, 219–33.

Schucany, W. R. and Woodward, W. A. (1977) Adjusting the degrees of freedom for the jack-knife. *Commun. Stat.*, **6**, 439–42.

Shaw, K. M., Lambshead, P. J. D. and Platt, H. M. (1983) Detection of pollution induced disturbance in marine benthic assemblages with special reference to nematodes. *Mar. Ecol. Progr. Ser.*, **11**, 195–202.

Shubert, L. E. (1984) *Algae as Ecological Indicators*, Academic Press, London.

Simberloff, D. (1972) Properties of rarefaction diversity measurements. *Amer. Nat.* **106**, 414–15.

Simberloff, D. (1986) Design of nature reserves. In *Wildlife Conservation Evaluation* (ed. M. B. Usher), Chapman and Hall, London, pp. 316–37.

Simberloff, D. and Abele, L. G. (1982) Refuge design and island biogeographic theory: effects of fragmentation. *Amer. Nat.*, **120**, 41–50.

Simberloff, D. and Gotelli, N. (1984) Effects of insularisation on plant species richness in the prairie-forest ecotone. *Biol. Conserv.*, **29**, 27–46.

Simpson, E. H. (1949) Measurement of diversity. *Nature*, **163**, 688.

Slocomb, J., Stauffer, B. and Dickson, K. L. (1977) On fitting the truncated lognormal distribution to species-abundance data using maximum likelihood estimation. *Ecology*, **58**, 693–6.

Smith, B. (1986) Evaluation of different similarity indices applied to data from the Rothamsted insect survey. Unpublished MSc Thesis, University of York.

Sokal, R. R. and Rohlf, F. (1981) *Biometry*, Freeman, San Francisco.

Soulé, M. E. (1983) What do we really know about extinction? In *Genetics and Conservation* (eds C. M. Schonewald-Cox, S. M. Chambers, B. MacBryde and L. Thomas), Benjamin/Cummings, Menlo Park, CA, pp. 111–24.

Soulé, M. E. and Wilcox, B. A. (1980) *Conservation Biology: an Evolutionary–Ecological Perspective*, Sinauer, Sunderland, MA.

Southwood, T. R. E. (1978) *Ecological Methods*, Chapman and Hall, London.

Southwood, T. R. E. (1988) The concept and nature of the community. In *Organisation of Communities: Past and Present* (eds), Blackwell, Oxford.

Southwood, T. R. E., Brown, V. K. and Reader, P. M. (1979) The relationship of plant and insect diversities in succession. *Biol. J. Linn. Soc.*, **12**, 327–48.

Southwood, T. R. E. and Kennedy, C. E. J. (1983) Trees as islands. *Oikos*, **41**, 359–71.

Spellerberg, I. F. (1981) *Ecological Evaluation for Conservation*, Arnold, London.

Stoermer, E. F. (1984) Qualitative characteristics of phytoplankton assemblages. In *Algae as Ecological Indicators* (ed. L. E. Shubert), Academic Press, London, pp. 49–67.

Sugihara, G. (1980) Minimal community structure: an explanation of species abundance patterns. *Amer. Nat.*, **116**, 770–87.

Taylor, L. R. (1978) Bates, Williams, Hutchinson – a variety of diversities. In *Diversity of Insect Faunas: 9th Symposium of the Royal Entomological Society* (eds L. A. Mound and N. Warloff), Blackwell, Oxford, pp. 1–18.

Taylor, L. R. (1986) Synoptic dynamics, migration and the Rothamsted Insect Survey. *J. Anim. Ecol.*, **55**, 1–38.

Taylor, L. R. and French, R. A. (1974) Effects of light trap design and illumination on samples of moths in an English woodland. *Bull. Entomol. Res.*, **63**, 583–94.

Taylor, L. R., Kempton, R. A. and Woiwod, I. P. (1976) Diversity statistics and the log-series model. *J. Anim. Ecol.*, **45**, 255–71.

Terborgh, J. (1975) Faunal equilibria and the design of wildlife preserves. In *Tropical Ecological Systems: Trends in Terrestrial and Aquatic Research* (eds F. Golley and E. Medina), Springer Verlag, New York, pp. 369–80.

Terborgh, J. (1977) Bird species diversity on an Andean elevational gradient. *Ecology*, **58**, 1007–19.

Thomas, M. R. and Shattock, R. C. (1986) Filamentous fungal associations in the phylloplane of *Lolium perenne*. *Trans. Brit. Mycol. Soc.*, **87**, 255–68.

Thoreau, H. D. (1860) The succession of forest trees. In *Excursions* (1863), Houghton and Mifflin, Boston.

Thorman, S. (1982) Niche dynamics and resource partitioning in a fish guild inhabiting a shallow estuary on the Swedish West Coast. *Oikos*, **39**, 32–9.

Tilman, D. (1982) *Resource Competition and Community Structure*, Princeton University Press, Princeton.

Tomascik, T. and Sander, F. (1987) Effects of eutrophication on reef building corals. II. Structure of scleractinian coral communities on fringing reefs, Barbados, West Indies. *Mar. Biol.*, **94**, 53–75.

Tukey, J. (1958) Bias and confidence in not quite large samples (abstract). *Ann. Math. Stat.*, **29**, 614.

Ugland, K. I. and Gray, J. S. (1982) Lognormal distributions and the concept of community equilibrium. *Oikos,* **39**, 171–8.

Usher, M. B. (1983) Species diversity: a comment on a paper by W. B. Yapp. *Field Stud.,* **5**, 825–32.

Usher, M. B. (1985) Implications of species-area relationships for wildlife conservation. *J. Environm. Manag.,* **21**, 181–91.

Usher, M. B. (1986) Wildlife conservation evaluation: attributes, criteria and values. In *Wildlife Conservation Evaluation* (ed. M. B. Usher), Chapman and Hall, London, pp. 3–44.

van der Ploeg, S. W. F. (1986) Wildlife conservation evaluation in the Netherlands: a controversial issue in a small country. In *Wildlife Conservation Evaluation* (ed. M. B. Usher), Chapman and Hall, London, pp. 162–80.

Vida, G. (1978) Genetic diversity and environmental future, *Environm. Conserv.,* **5**, 127–32.

Webb, D. J. (1974) The statistics of relative abundance and diversity. *J. Theor. Biol.,* **43**, 277–92.

Whittaker, R. H. (1960) Vegetation of the Siskiyou Mountains, Oregon and California. *Ecol. Monogr.,* **30**, 279–338.

Whittaker, R. H. (1965) Dominance and diversity in land plant communities. *Science,* **147**, 250–60.

Whittaker, R. H. (1970) *Communities and Ecosystems,* Macmillan, New York.

Whittaker, R. H. (1972) Evolution and measurement of species diversity. *Taxon,* **21**, 213–51.

Whittaker, R. H. (1977) Evolution of species diversity in land communities. In *Evolutionary Biology,* Vol. 10 (eds M. K. Hecht, W. C. Steere and B. Wallace), Plenum, New York, pp. 1–67.

Williams, C. B. (1964) *Patterns in the Balance of Nature and Related Problems in Quantitative Ecology,* Academic Press, London.

Williamson, M. (1981) *Island Populations,* Oxford University Press, Oxford.

Williamson, M. H. (1973) Species diversity in ecological communities. In *The Mathematical Theory of the Dynamics of Biological Populations* (eds M. S. Bartlett and R. W. Hiorns), Academic Press, London, pp. 325–35.

Wilson, M. V. and Shmida, A. (1984) Measuring beta diversity with presence-absence data. *J. Ecol.,* **72**, 1055–64.

Wilson, M. V. and Mohler, C. L. (1983) Measuring compositional change along gradients. *Vegetatio,* **54**, 129–41.

Wilson, E. O. and Willis, E. O. (1975) Applied biogeography. In *Ecology and Evolution of Communities* (eds M. L. Cody and J. M. Diamond), Harvard University Press, Cambridge, MA, pp. 522–34.

Wolda, H. (1981) Similarity indices, sample size and diversity. *Oecologia,* **50**, 296–302.

Wolda, H. (1983) Diversity, diversity indices and tropical cockroaches. *Oecologia,* **58**, 290–8.

Wu, R. S. S. (1982) Periodic defaunation and recovery in a subtropical epibenthic community, in relation to organic pollution. *J. Exp. Mar. Biol. Ecol.,* **64**, 253–69.

Yapp, W. B. (1979) Specific diversity in woodland birds. *Field Stud.,* **5**, 45–58.

Zahl, S. (1977) Jack-knifing an index of diversity. *Ecology,* **58**, 907–13.

Zipf, G. K. (1965) *Human Behaviour and the Principle of Least Effort,* Hafner, New York.

Worked examples

1 Rarefaction

It is not always possible to ensure that sample sizes are equal. Rarefaction is one way of coping with this difficulty. It is a method of working out the number of species that would be expected in samples of a standard size. The technique was devised by Sanders (1968) but this example uses the Hurlbert's (1971) unbiased version of the formula. The method has a number of drawbacks. First the calculations involve many factorials and are tedious. Secondly, as indicated in Chapter 2, rarefaction leads to a great loss of information. The formula is

$$E(S) = \Sigma\left\{1 - \left[\binom{N-N_i}{n} \Big/ \binom{N}{n}\right]\right\}$$

where $E(S)$ = the expected number of species in the rarefied sample
n = standardized sample size
N = the total number of individuals recorded in the sample to be rarefied
N_i = the number of individuals in the ith species in the sample to be rarefied.

The simplest approach is to take the number of individuals in the smallest sample as the standardized sample size. This minimizes the (not inconsiderable) calculations involved. Data from two moth traps are used in this example. (Strictly speaking this is not a true application of rarefaction since sampling effort, that is the length of time over which the traps were set, was equal. The small numbers however serve to illustrate the calculations.) Twenty-three individuals were collected in the first trap but only 13 were found in the second. How many species would we have expected in Trap A if it too had contained 13 individuals? The answer, as the following calculations illustrate, is 6.6 species.

Species	*Trap A*	*Trap B*
July highflyer	9	1
Dark arches	3	0
Silver Y	0	1
Coxcomb prominent	4	0
True lover's knot	2	0
Buff tip	1	0
Snout	1	1
Barred red	0	2
Swallow prominent	1	0
Antler	0	5
Large yellow underwing	1	3
Beautiful golden Y	1	0
Number of species (S)	9	6
Number of individuals (N)	23	13

1. The term $\binom{x}{y}$ is a 'combination' which is calculated as follows:

$$\binom{x}{y} = \frac{x!}{y!(x-y)!}$$

$x!$ is a factorial. For example $5! = 5 \times 4 \times 3 \times 2 \times 1 = 120$.

With these points in mind the computations can proceed.

2. The first step is to take each species abundance from Trap A and insert it in the formula.

$$\left\{1 - \left[\binom{N-N_i}{n} \Big/ \binom{N}{n}\right]\right\}$$

Thus, for the July highflier, which was represented by nine individuals, the calculations are

$$\left\{1 - \left[\left(\frac{14!}{13! \times 1!}\right) \Big/ \left(\frac{23!}{13! \times 10!}\right)\right]\right\} = \{1 - [14/1144066]\}$$

$$= 1 - 0.00 = 1.00.$$

The result for each species is listed and summed to give the expected species number for Trap A.

N_i	
9	1.00
3	0.93
4	0.98
2	0.82
1	0.57
1	0.57
1	0.57
1	0.57
1	0.57
Expected number of species for Trap A $E(S) =$	6.58

Species richness measures

Two simple species richness measures are the Margalef and Menhinick indices. Chapter 2 gives details.

They are calculated from the following formulae:

Margalef's index $D_{Mg} = (S-1)/\ln N$
Menhinick's index $D_{Mn} = S/\sqrt{N}$

The diversity of the two moth traps listed above would be:

	Trap A	*Trap B*
Margalef	2.55	1.95
Menhinick	1.88	1.66

2 Geometric series

The geometric series is most commonly applied to species-poor assemblages. The basic assumption is that the dominant species will use proportion k of some limiting resource, the second most dominant species will take proportion k of the remainder and so on until all species have been accounted for. The abundance of each species is assumed to be equivalent to the proportion of the resource it uses. In a geometric series the abundances of species, ranked from most abundant to least abundant, are therefore

$$n_i = NC_k k(1-k)^{i-1}$$

where k = the proportion of the available niche space or resource that each species occupies
n_i = the number of individuals in the ith species
N = the total number of individuals, and
$C_k = [1-(1-k)^S]^{-1}$, and is a constant which ensures that $\Sigma n_i = N$

This example tests whether the Collembola (springtails) in the soil of a conifer plantation follow a geometric series. A Tullgren Funnel was used to extract the Collembola from 10 soil cores. The number of species and individuals obtained is listed below.

Collembola

Species	*Individuals*
Folsomia sp.	370
Isotoma sp. A	210
Isotoma sp. B	120
Entomobrya sp.	66
Isotomiella sp.	35
Tetrodontophora sp.	31
Willemia sp.	15
Isotumurus sp.	9
Orchesella sp.	3
Lepidocyrtus sp.	2
Willowsia sp.	1

Number of species (S) = 11
Number of individuals (N) = 862

1. In order to fit the geometric series it is necessary to begin by estimating the constant k. This is done by iterating the following equation (see May, 1975 for details)

$$N_{min}/N = [k/(1-k)]\,[(1-k)^s]/[1-(1-k)^s]$$

where N_{min} is the number of individuals in the least abundant species. In this example $N_{min}/N = 0.00116$. Solving this equation requires trying successive values of k until the two sides of the equation balance.

For example try $k = 0.42$ $[k/(1-k)]\,[(1-k)^s]/[1-(1-k)^s] = 0.00181$
$k = 0.44$ $= 0.00134$
$k = 0.45$ $= 0.00114$
$k = 0.449$ $= 0.00116$

2. With k estimated as 0.449 it is now possible to obtain the value of C_k:

$$C_k = [1-(1-k)^s]^{-1} = [1-(1-0.449)^{11}]^{-1} = 1.001432$$

and calculate the expected number of individuals for each of the 11 species. Thus for the most abundant species

$$n_i = NC_k k(1-k)^{i-1} = 862 \times 1.001432 \times 0.449 \times (1-0.449)^0 = 387.6.$$

3. When this process has been repeated for each successive species the observed and expected values can be compared using a χ^2 goodness of fit (GOF) test. χ^2 is Σ (observed − expected)^2/expected. χ^2 tables show that there is no significant difference between the observed and expected abundances of each springtail species with a probability of $P > 0.30$ ($df = s - 1 = 10$). Thus we can conclude that the Collembola follow a geometric series. Linear regression may also be used to measure goodness of fit. The simplest approach of all is to compile a rank abundance plot (see for example Figure 4) and examine it to see whether all points lie on a straight line.

Species	*Observed*	*Expected*	χ^2
Folsomia sp.	370	387.6	0.80
Isotoma sp. A	210	213.8	0.07
Isotoma sp. B	120	117.8	0.04
Entomobrya sp.	66	64.5	0.03
Isotomiella sp.	35	35.5	0.01
Tetrodontophora sp.	31	19.8	6.34
Willemia sp.	15	10.7	1.73
Isotumurus sp.	9	6.2	1.26
Orchesella sp.	3	3.3	0.03
Lepidocyrtus sp.	2	1.8	0.02
Willowsia sp.	1	1.0	0.00
	$\Sigma n_i = 862$	862	$\Sigma\chi^2$ 10.33

3 The log series

Thomas and Shattock (1986) were interested in the filamentous fungal associations in the phylloplane of the grass *Lolium perenne*. As part of their study they assembled a list giving the total relative abundance of species on the leaves of *L. perenne*. It is these data that are used to illustrate the calculations involved when fitting the log series. A full discussion of the log series is to be found on page 17 in Chapter 2. This example concerns itself simply with the mechanics of the calculations involved.

Species	*Abundance* (n_i)
Cladosporium	1988
Drechslera andersenii	1358
Phoma	1042
Epicoccum	994
Alternaria	607
Leptosphaeria	533
Fusarium	324
D. siccans	299
Ascochyta	150
Acremonium	136
Streptomyces	125
Dinemasporium	101
Rhynchosporium	43
Stemphylium	40
Botrytis	24
Septoria	16
Cheilaria	14
Dendryphion	12
Humicola	8
Chrysosporium	7
Gonatbotrys	7
Torula	7
Rhizopus	6
Acremoniella	3
Erysiphe	3
Papulaspora	3
Puccinia	3
Stachybotrys	3
Arthrobotrys	1
Chaetomium	1
Colletotrichium	1
Periconia	1
Pleospora	1

Number of species (S) = 33
Number of individuals (N) = 7861

1. It is often useful to start by drawing a rank abundance plot. See for example Figures 2.4 and 2.6.
2. Put the observed abundances into abundance classes. For reasons of comparability it is best to use the same abundance classes as those adopted when fitting the log normal and broken stick distributions (see examples 4 and 5). In this case classes in $\log_2$ (that is octaves or doublings of species abundances) are chosen. Adding 0.5 to the upper boundary of each class makes it straightforward to unambiguously assign observed species abundances to each class. Thus in the table below there are five species with an abundance of one or two individuals, a further five species with an abundance of three or four individuals, and so on.

Class	*Upper boundary*	*Number of species observed*
1	2.5	5
2	4.5	5
3	8.5	5
4	16.5	3
5	32.5	1
6	64.5	2
7	128.5	2
8	256.5	2
9	512.5	2
10	1024.5	3
11	∞	3

Total number of species $(S) = 33$

3. The log series takes the form:

$$\alpha x, \frac{\alpha x^2}{2}, \frac{\alpha x^3}{3} \cdots \frac{\alpha x^n}{n} \qquad \text{(see equation 2.5, p. 18)}$$

with αx being the number of species with one individual, $\alpha x^2/2$ the number of species with two individuals, etc.

To fit the series it is necessary to calculate how many species are expected to have one individual, two individuals and so on. These expected abundances are then put into the same abundance classes used for the observed distribution and a goodness of fit test is used to compare the two distributions. The total number of species in the observed and expected distributions is of course identical.

The two parameters needed to fit the series are x and α. x is estimated by iterating the following term

$$S/N = [(1-x)/x]\,[-\ln(1-x)]$$

where S = total number of species and N = total number of individuals. x is usually greater than 0.9 and always <1.0. In cases where the ratio $N/S > 20$, x will be >0.99. In this example $N/S = 7861/33 = 238.21$. A few calculations on a hand calculator will quickly produce the correct value of x.

$$S/N = 0.00420$$

try $x = 0.995$	$[(1-0.995)/0.995]\,[-\ln(1-0.995)]$	
$x = 0.995$		$S/N = 0.02665$
try $x = 0.999$		$S/N = 0.00691$
try $x = 0.9999$		$S/N = 0.00092$
try $x = 0.9995$		$S/N = 0.00380$
try $x = 0.9994$		$S/N = 0.00445$
try $x = 0.99945$		$S/N = 0.00413$
try $x = 0.99944$		$S/N = 0.00420$

The correct value of x is therefore 0.99944. Once x has been obtained it is simple to calculate α using the equation

$$\alpha = \frac{N(1-x)}{x} = \frac{7861 \times (1-0.99944)}{0.99944} = 4.4046$$

α is an index of diversity.

When α and x have been obtained the number of species expected to have 1, 2, 3, . . . n individuals can be calculated. This is illustrated below for the first four abundance classes.

Number of individuals		*Number of species expected*	Σ
1	αx	4.4021	6.602
2	$\alpha x^2/2$	2.1998	
3	$\alpha x^3/3$	1.4657	2,564
2	$\alpha x^4/4$	1.0987	
5	$\alpha x^5/5$	0.8785	
6	$\alpha x^6/6$	0.7136	2.785
7	$\alpha x^7/7$	0.6286	
8	$\alpha x^8/8$	0.5481	
9	$\alpha x^9/9$	0.4869	
10	$\alpha x^{10}/10$	0.4390	
11	$\alpha x^{11}/11$	0.3980	
12	$\alpha x^{12}/12$	0.3648	
13	$\alpha x^{13}/13$	0.3364	2.900
14	$\alpha x^{14}/14$	0.3122	
15	$\alpha x^{15}/15$	0.2912	
16	$\alpha x^{16}/16$	0.2728	
etc.			

4. The next stage is to compile a table giving the number of expected and observed species in each abundance class and compare the two distributions using a goodness of fit (GOF) test. χ^2 is one commonly used test. For each class calculate χ^2 as shown.

$$\chi^2 = (\text{observed} - \text{expected})^2/\text{expected}.$$

For example, in class 1, $\chi^2 = (5 - 6.6)^2/6.6 = 0.39$. Finally sum this column to obtain the overall goodness of fit, $\Sigma\chi^2$. Check the obtained value in χ^2 tables (Appendix 1) using number of classes − 1 degrees of freedom. In this case $\Sigma\chi^2 = 7.21$. With 10 degrees of freedom the value of χ^2 for $P = 0.05$ is 18.307. For $P = 0.70$ it is 7.267. We can therefore conclude that there is no significant difference between the observed and expected distributions with a probability of $P > 0.70$.

Class	*Upper boundary*	*Observed*	*Expected*	χ^2
1	2.5	5	6.6	0.39
2	4.5	5	2.6	2.22
3	8.5	5	2.8	1.73
4	16.5	3	2.9	0.00
5	32.5	1	2.9	1.24
6	64.5	2	2.9	0.28
7	128.5	2	2.9	0.28
8	256.5	2	2.7	0.18
9	512.5	2	2.5	0.10
10	1024.5	3	2.2	0.29
11	∞	3	2.0	0.50
Number of species		33	33	$\Sigma\chi^2 = 7.21$

If χ^2 is calculated when the number of expected species is small (<1.0) the resultant value of χ^2 can be extremely large. In such cases it is best to combine the observed number of species in two or more adjacent classes and compare this with the combined number of expected species in the same two classes. The degrees of freedom should be reduced accordingly.

Source: Thomas, M. R. and Shattock, R. C. (1986) Filamentous fungal associations in the phylloplane of *Lolium perenne*. *Trans. Brit. Mycol. Soc.*, **87**, 255–68.

4 The truncated log normal

Fitting a conventional log normal is straightforward and standard statistics texts will provide details. As Chapter 2 pointed out most log normals that are encountered in investigation of species abundance data are of the truncated variety. This example illustrates Pielou's (1975) method of fitting a truncated log normal. The data used to illustrate the log series in Example 3 are also employed in this example.

Species	*Abundance* (n_i)
Cladosporium	1988
Drechslera andersenii	1358
Phoma	1042
Epicoccum	994
Alternaria	607
Leptosphaeria	533
Fusarium	324
D. siccans	299
Ascochyta	150
Acremonium	136
Streptomyces	125
Dinemasporium	101
Rhynchosporium	43
Stemphylium	40
Botrytis	24
Septoria	16
Cheilaria	14
Dendryphion	12
Humicola	8
Chrysosporium	7
Gonatbotrys	7
Torula	7
Rhizopus	6
Acremoniella	3
Erysiphe	3
Papulaspora	3
Puccinia	3
Stachybotrys	3
Arthrobotrys	1
Chaetomium	1
Colletotrichium	1
Periconia	1
Pleospora	1

Number of species (S) = 33
Number of individuals (N) = 7861

1. Since this is a log normal distribution the first step is to log each of the species abundances ($x = \log_{10} n_i$) and obtain the observed mean and variance. The mean and variance are calculated in the normal way ($\bar{x} = \Sigma x/S$ and $\sigma^2 = \Sigma(x-\bar{x})^2/S$). In this example $\bar{x} = 1.392$ and $\sigma^2 = 1.114$.
2. Next calculate $\gamma = \sigma^2/(x-x_0)^2$ where $x_0 = -0.30103$. $\gamma = 0.389$.
3. Use Appendix 3 (Cohen's, 1961; Table 1) to get the 'auxiliary estimation function' θ for this value of $\gamma = 0.389$. Here $\theta = 0.2429$.
4. Obtain the estimates of μ_x and V_x of the mean and variance of x using the equations

$$\mu_x = \bar{x} - \theta(\bar{x} - x_0) \text{ and } V_x = \sigma^2 + \theta(\bar{x} - x_0)^2.$$

Thus $\mu_x = 0.98$ and $V_x = 1.823$.
5. Find what is termed the 'standardized normal variate' z_0, which corresponds to the truncation point x_0, from the equation $z_0 = (x_0 - \mu_x)/\sqrt{V_x}$. Here $z_0 = -0.949$.
6. Use tables that give the area under the normal curve to find the value p_0. Here $p_0 = 0.171$. This represents the unsampled species in the community, that is, the ones to the left of the veil line.
7. The value of p_0 can be used to obtain the total number of species in the community S^*. The equation $S^* = S/(1-p_0)$ is employed. Therefore $S^* = 33/(1-0.171) = 39.8$.
8. With these values obtained it is now possible to estimate the number of species expected in each class. To do this it is helpful to construct a table with the following columns:
(a) the upper class boundary (for comparability the abundance classes are the same as those used in the log series and broken stick distributions (see examples 3 and 5);
(b) the upper class boundary converted to $\log_{10}$ (for class 3 for example $\log_{10}$ of 8.5 is 0.929);
(c) the standardized form of these logged upper class boundaries, that is $[b - \mu_x]/\sqrt{V_x}$ (in class 3 the value will be -0.037); and
(d) the cumulative number of species expected.

Each successive class represents another step across the log normal distribution and therefore the area accounted for is equivalent to the number of species expected. To obtain the values in this column take each value in (c), look it up in the same tables used in step 6, and multiply the result by S^*, the expected total number of species. Thus for class 3 the result will be $39.8 \times 0.484 = 19.27$. Differences between successive entries provide the expected number of species in each class.

	Upper boundary (a)	Log_{10} UB (b)	Standardized UB (c)	Σ expected species (d)
	0.5	−0.301	−0.949	6.8
1	2.5	0.398	−0.431	13.22
2	4.5	0.653	−0.242	16.05
3	8.5	0.929	−0.037	19.27
4	16.5	1.217	0.176	22.65
5	32.5	1.512	0.394	25.97
6	64.5	1.810	0.614	29.06
7	128.5	2.109	0.836	31.76
8	256.5	2.409	1.059	34.02
9	512.5	2.710	1.281	35.81
10	1024.5	3.011	1.504	37.15
11	∞	∞	∞	39.78

9. Next, calculate λ, the log normal diversity statistic. This is obtained from the equation $\lambda = S^*/\sigma = 39.8/1.35 = 29.5$.
10. Finally, compare the observed and expected number of species using a χ^2 goodness of fit test. This procedure was illustrated for the log series in example 3. In this example χ^2 GOF$=7.53$. Eight degrees of freedom (that is df= classes − 3) are required. We can therefore conclude that these data, which were described by the log series, are also described by the truncated log normal at a probability of $P=0.50$.

Class	Upper boundary	Observed	Expected	χ^2
Behind veil line	0.5	–	6.8	–
1	2.5	5	6.4	0.3
2	4.5	5	2.8	1.7
3	8.5	5	3.2	1.0
4	16.5	3	3.4	0.0
5	32.5	1	3.3	1.6
6	64.5	2	3.1	0.4
7	128.5	2	2.7	0.2
8	256.5	2	2.3	0.0
9	512.5	2	1.8	0.0
10	1024.5	3	1.3	2.0
11	∞	3	2.6	0.1
				$\Sigma\chi^2$ 7.35

Source: Thomas, M. R. and Shattock, R. C. (1986) Filamentous fungal associations in the phylloplane of *Lolium perenne*. *Transactions of the British Mycological Society*, **87**, 255–86.

5 The broken stick

One method of fitting the broken stick model involves working out the expected number of individuals in the ith most abundant of S species (see Chapter 2 for details). An alternative approach, and the one adopted here, is to calculate the number of species expected in the abundance class with n individuals. This facilitates comparison between the resultant expected values and those obtained for the log series and truncated log normal. The broken stick is illustrated with data collected by Driscoll (1977). These data record the variety and abundance of species of birds occurring in wet sclerophyll forest in Australia.

Species	*Abundance*
Gang-gang cockatoo	103
Crimson rosella	115
Laughing kookaburra	13
Superb lyrebird	2
Striated thornbill	67
Brown thornbill	36
White-browed scrub-wren	51
Flame robin	8
Southern yellow robin	6
Grey fantail	61
Golden whistler	10
Grey shrike-thrush	21
Eastern whipbird	7
White-throated tree-creeper	65
Red-browed tree-creeper	4
Yellow-faced honeyeater	49
White-eared honeyeater	92
White-naped honeyeater	37
Noisy friarbird	16
Red-browed finch	6
Pied currawong	23
Raven	9
Rufous fantail	2
Satin flycatcher	6
Rufous whistler	5
Eastern shrike-tit	4
Eastern striated paradalote	1
Grey-breasted silvereye	3
Crescent honeyeater	1
Eastern spinebill	9
Black-backed magpie	2

Number of species (S) = 31
Total number of individuals (N) = 834

1. As with the log series and truncated log normal, the first step is to allocate the observed species to abundance classes. Log_2 classes are used in this example.
2. It is then necessary to calculate the number of species expected to have one individual, two individuals, etc.

This is done using the formula

$$S(n) = [S(S-1)/N]\,(1 - n/N)^{S-2}$$

where $S(n)$ is the number of species in the abundance class with n individuals. Therefore we would expect the following number of species to have one individual

$$[31 \times 30/834] \times (1 - 1/834)^{29} = 1.077$$

A table of $S(n)$ can now be constructed.

Number of individuals	*Number of species expected*	Σ
1	1.077	2.117
2	1.040	
3	1.004	1.974
2	0.970	
5	0.937	3.557
6	0.904	
7	0.873	
8	0.843	
9	0.814	5.776
10	0.786	
11	0.759	
12	0.732	
13	0.707	
14	0.683	
15	0.659	
16	0.636	

When this is complete the expected number of species are placed alongside the observed number of species in the $\log_2$ abundance classes and χ^2 GOF calculated as before.

Class	*Upper boundary*	*Observed*	*Expected*	χ^2
1	2.5	5	2.1	3.93
2	4.5	3	2.0	0.53
3	8.5	6	3.6	1.68
4	16.5	5	5.8	0.10
5	32.5	2	7.6	4.14
6	64.5	5	6.6	0.40
7	128.5	5	2.6	2.31
8	256.5	0	0.2	0.20
9	512.5	0	0.0	0.00
10	1024.5	0	0.0	0.00
11	∞	0	0.6	0.55
Number of species		31	31	$\Sigma\chi^2 = 13.84$

χ^2 tables show that with 10 degrees of freedom (classes − 1) the probability of the expected and observed distributions being significantly different is $P > 0.10$. Since the last four rows in the table are effectively empty it is more conservative to reduce the degrees of freedom to six. When this is done the probability of the two distributions differing becomes $P < 0.05$. We are clearly dealing here with a species abundance distribution which is approaching a broken stick distribution.

Source: Driscoll, P. V. (1977) Comparison of bird counts from pine forests and indigenous vegetation. *Australian Wildlife Research*, **4**, 281–8.

6 The Q statistic

The Q statistic is a measure of the inter-quartile slope of the cumulative species abundance curve (see Figure 2.14). It is a robust and useful diversity measure which does not require the fitting of a species abundance model. The calculations involved are illustrated using data collected on the ground flora in Breen oakwood, Northern Ireland. The vegetation was sampled using 50 randomly placed point quadrats. Abundances are the number of hits, or points, per species.

Species	*Abundance*
Potentilla erecta	20
Oxalis acetosella	63
Anthoxanthum odoratum	33
Deschampsia flexuosa	140
Luzula sylvatica	170
Calluna vulgaris	7
Vaccinium myrtillus	133
Blechnum spicant	10
Polytrichum formosum	38
Thuidium tamariscinum	15
Dicranum majus	11
Molinia caerula	52
Holcus lanatus	37
Juncus effusus	13
Pteridium aquilinum	29
Poa trivialis	2
Gallium saxatile	3
Rhytidiadelphus loreus	4
Rhytidiadelphus triquetrus	33
Holcus mollis	6
Sphagnum acutifolium	15
Sphagnum palustre	8
Hypnum cupressiforme	6
Rhytidiadelphus squarrosus	9
Agrostis tenuis	14
Carex flexuosa	3
Dryopteris dilatata	4
Mnium hornum	3
Pseudoscleropodium purum	3

Number of species (S) = 29
Number of individuals (N) = 877

1. The first step when calculating the Q statistic is to assemble a table showing the cumulative number of species against abundance (see below) and work out the position of the lower and upper quartiles, i.e. the points at which 25% and 75% of the species lie. One quarter of 29 species is 7.25 while three-quarters of 29 is 21.75. The lower quartile (*R*1) should be chosen so that the cumulative number of species in the class in which it occurs is greater than, or equal to, 25% of the total number of species. Likewise, the upper quartile, *R*2, falls in the class with greater than or equal to 75% of the total number of species. In this example *R*1 occurs when the cumulative number of species reaches 8 and *R*2 is found at the point where the cumulative number of species is 22. The exact choice of *R*1 and *R*2 is relatively unimportant. Equation 2.16 (page 34) is the formal way of expressing the choice of quartiles.

	Number of individuals	*Number of species*	Σ *Number of species*	
	2	1	1	
	3	5	6	
R	**4**	**3**	**8**	**Lower quartile *R*1**
	6	2	10	
	7	2	12	
	9	2	14	
	11	1	15	
	14	1	16	
	15	2	18	
	20	1	19	
	29	1	20	
	33	1	21	
***R*2**	**34**	**1**	**22**	**Upper quartile *R*2**
	36	1	23	
	37	1	24	
	53	1	25	
	57	1	26	
	138	1	27	
	146	1	28	
	170	1	29	

2. Once the quartiles are selected it is simple to calculate Q using the equation

$$Q = \frac{\frac{1}{2}n_{R1} + \Sigma n_r + \frac{1}{2}n_{R2}}{\ln(R2/R1)}$$

where $\frac{1}{2}n_{R1}$ = half of the number of species in the class where the lower quartile falls;

Σn_r = the total number of species between the quartiles;

$\frac{1}{2}n_{R2}$ = half of the number of species in the class where the upper quartile falls;

$R1$ = the number of individuals in the class with the lower quartile;

$R2$ = the number of individuals in the class with the upper quartile.

Therefore in this example

$$Q = \frac{1.5 + 13 + 0.5}{\ln(34/4)} = 7.01$$

7 Shannon diversity index

Batten (1976) recorded bird species richness and abundance in a number of native woodlands and conifer plantations in Killarney, Ireland. The aim of the study was partly to determine whether conifer plantations are impoverished relative to the endemic woodlands. In this example the diversity of two of the woodlands, Derrycunihy oakwood (area 10.75 ha) and a Norway spruce plot (area 11 ha), is estimated using Shannon's diversity index. A *t* test is used to test for differences in the diversity of the two sites.

Derrycunihy oakwood

Species	*Number of territories*
Chaffinch	35
Robin	26
Blue tit	25
Goldcrest	21
Wren	16
Coal tit	11
Spotted flycatcher	6
Tree-creeper	5
Siskin	3
Blackbird	3
Great tit	3
Long-tailed tit	3
Woodpigeon	3
Hooded crow	2
Woodcock	2
Song thrush	2
Redstart	1
Mistle thrush	1
Dunnock	1
Sparrowhawk	1

Number of species (S) = 20
Number of territories (N) = 170

Norway spruce	
Species	*Number of territories*
Goldcrest	65
Robin	30
Chaffinch	30
Wren	20
Blackbird	14
Coal tit	11
Woodpigeon	9
Song thrush	5
Tree creeper	4
Blue tit	3
Long-tailed tit	3
Siskin	2
Redpoll	1
Crow	1
Number of species (S) = 14	
Number of territories (N) = 198	

1. The formula for calculating the Shannon diversity index is

$$H' = -\Sigma p_i \ln p_i$$

where p_i, the proportional abundance of the ith species $= (n_i/N)$.

Thus the first step when calculating the index by hand is to draw up a table giving values of p_i and $p_i \ln p_i$. In cases where t test is also being used it is convenient to add a further column to the table giving values of $p_i\ (\ln p_i)^2$.

The tables for the two woodlands are shown below and overleaf.

Derrycunihy oakwood

Territories	p_i	$p_i \ln p_i$	$p_i(\ln p_i)^2$
35	0.206	−0.325	0.514
26	0.153	−0.287	0.539
25	0.147	−0.282	0.540
21	0.124	−0.258	0.540
16	0.094	−0.222	0.526
11	0.065	−0.177	0.485
6	0.035	−0.118	0.395
5	0.029	−0.104	0.366
3	0.018	−0.071	0.288
3	0.018	−0.071	0.288
3	0.018	−0.071	0.288
3	0.018	−0.071	0.288
3	0.018	−0.071	0.288
2	0.012	−0.052	0.232
2	0.012	−0.052	0.232
2	0.012	−0.052	0.232
1	0.006	−0.030	0.155
1	0.006	−0.030	0.155
1	0.006	−0.030	0.155
1	0.006	−0.030	0.155
1	0.006	−0.030	0.155
Σ170	1.000	−2.404	6.661

Norway spruce			
Territories	p_i	$p_i \ln p_i$	$p_i(\ln p_i)^2$
65	0.328	−0.366	0.407
30	0.152	−0.286	0.540
30	0.152	−0.286	0.540
20	0.101	−0.232	0.531
14	0.071	−0.187	0.496
11	0.056	−0.161	0.464
9	0.054	−0.141	0.434
5	0.025	−0.093	0.342
4	0.020	−0.079	0.308
3	0.015	−0.063	0.266
3	0.015	−0.063	0.266
2	0.010	−0.046	0.213
1	0.005	−0.027	0.141
1	0.005	−0.027	0.141
Σ198	1.000	−2.056	5.089

2. Once these tables are assembled it is simple to proceed with the remainder of the calculations. The diversity of the oakwood is $H'=2.404$ while the diversity of the spruce plantation is $H'=2.056$. These values represent the sum of the $p_i \ln p_i$ column. The formula for the Shannon index commences with a minus sign to cancel out the negative created by taking logs of proportions.

The evenness of the two woodlands can now be calculated using the formula

$$E=H'/\ln S$$

Oakwood evenness $=2.404/\ln 20=0.8025$
Spruce plantation evenness $=2.056/\ln 14=0.7791$

3. The variance in diversity of the two woodlands may be estimated using the formula

$$\text{Var } H'=\frac{\Sigma p_i(\ln p_i)^2-(\Sigma p_i \ln p_i)^2}{N}-\frac{S-1}{2N^2}$$

Thus

$$\text{Var } H' \text{ (oakwood)}=\frac{6.661-5.779}{170}-\frac{19}{340^2}=0.00502$$

and

$$\text{Var } H' \text{ (spruce)}=\frac{5.089-4.227}{198}-\frac{13}{396^2}=0.00427$$

4. A t test allows the diversities of the two woodlands to be compared. The appropriate formula is

$$t = \frac{H'_1 - H'_2}{(\text{Var } H'_1 + \text{Var } H'_2)^{1/2}}$$

where H'_1 is the diversity of site 1 and Var H'_1 is its variance.
In this example

$$t = \frac{2.404 - 2.056}{(0.00502 + 0.00427)^{1/2}} = 3.611$$

The requisite degrees of freedom must also be calculated. The formula required is

$$df = \frac{(\text{Var } H'_1 + \text{Var } H'_2)^2}{[(\text{Var } H'_1)^2/N_1] + [(\text{Var } H'_2)^2/N_2]}$$

where N_1 is the number of individuals (territories) in site 1.
Therefore

$$df = \frac{(0.00502 + 0.00427)^2}{(0.00502^2/170) + (0.00427^2/198)} = 360$$

t tables will quickly reveal that the two woods are highly significantly different ($P < 0.001$) in terms of the diversity of bird territories occurring in them. The native oakwood is thus more diverse than the spruce plantation.

Source: Batten, L. A. (1976) Bird communities of some Killarney woodlands. *Proc. Roy. Irish Acad.*, **76**, 285–313.

8 Brillouin index

The Brillouin index should be used instead of the Shannon index when the diversity of non-random samples or collections is being estimated. For instance, light traps produce biased samples of Macrolepidoptera since not all species are equally attracted by light. The Brillouin index is used here to calculate the diversity of moths collected in a portable light trap left out overnight in early summer in Banagher oakwood, Northern Ireland.

Species	*Number of individuals*	$\ln n_i!$
Small angle shade	17	33.505
July highflyer	15	27.899
Dark arches	11	17.502
Beautiful golden Y	4	3.178
Gallium carpet	4	3.178
Marbled carpet	3	1.792
Angle shade	3	1.792
Snout	3	1.792
Scalloped oak	2	0.693
Small yellow underwing	2	0.693
Purple clay	1	0
Silver ground carpet	1	0
Riband wave	1	0
Total number of individuals (N)	$N=\Sigma n_i=67$	$\Sigma(\ln n_i!)=92.024$
Total number of species (S) $=13$		

1. The data table is presented in the usual way to show the number of individuals (n_i) in each species. There is however an additional column giving values of $\ln n_i!$ This is because the equation for the Brillouin index is

$$HB=\frac{\ln N!-\Sigma \ln n_i!}{N}$$

The symbol ! signifies a factorial. For instance 4! is $4\times3\times2\times1=24$. ln 4! is therefore $\ln 24=3.178$.

In this example

$$HB=\frac{67!-92.024}{67}=1.876$$

2. As diversity is being calculated for a collection there is no significance test. Each value of the index is automatically significantly different from every

other. It is however possible to calculate an additional evenness measure using the equation

$$E = \frac{HB}{HB_{max}}$$

where HB_{max} is given by

$$HB_{max} = 1/N \ln\left(\frac{N!}{\{[N/S]!\}^{s-r}\{([N/S]+1)!\}^{r}}\right)$$

and

$$[N/S] = \text{the integer of } N/S$$
$$r = N - S[N/S]$$

In this example $N/S = 67/13 = 5.15$

Therefore

$$[N/S] = 5 \text{ and } r = 67 - 13 \times 5 = 2$$
$$[N/S]! = 5! = 120$$
$$120^{s-r} = 120^{11}$$
$$([N/S]+1)! = 6! = 720$$
$$720^{r} = 720^{2}$$

Putting these calculations together we get

$$HB_{max} = 1/67 \ln\left(\frac{67!}{120^{11} \times 720^{2}}\right)$$
$$= 2.268$$

Evenness can now be calculated

$$E = 1.876/2.268 = 0.827$$

It is clear from the above example that the use of factorials in the equations quickly produce huge numbers. These may exceed the capacity of pocket calculators. It is however worth noting that many sets of statistical tables include a table giving values of ln $x!$ or log $x!$. There is no reason why the index should be calculated using natural logs although they have been employed in this example.

9 Simpson's index

The calculation of Simpson's index is illustrated using a data set which lists the total numbers of trees in an 8 acre (33.3 ha) study plot in an upland Ozark forest in Arkansas, USA. These data were collected by James and Shugart (1970) during an investigation of the habitats of breeding birds in Arkansas.

Species	*Number of individuals* (n_i)
Ulmus alata	752
Quercus stellata	276
Quercus velutina	194
Cercis canadensis	126
Celtis occidentalis	121
Ulmus americana	97
Ulmus rubra	95
Fraxinus americana	83
Morus rubra	72
Quercus muchlenbergii	44
Juniperus virginiana	39
Carya cordiformis	16
Cornus florida	15
Maclura pomifera	13
Gleditsia tricanthos	9
Quercus alba	9
Carya texana	9
Prunus americana	8
Prunus serotina	7
Juglans nigra	4
Ligustrum sp.	2
Crataegus sp.	2
Diospyros virginiana	1
Viburnum rufidulum	1
Quercus falcata	1

Number of species (S) = 25
Number of individuals (N) = 1996

The equation used to calculate Simpson's index is

$$D = \sum \frac{(n_i(n_i - 1))}{(N(N-1))}$$

where n_i = the number of individuals in the ith species, and N = the total number of individuals.

Therefore in this data set the calculations will be

$$\{(752 \times 751)/(1996 \times 1995) + (276 \times 275)/1996 \times 1995) + \cdots + (1 \times 0)/(1996 \times 1995)\} = 0.187$$

The reciprocal form of Simpson's index is usually adopted. This ensures that the value of the index increases with increasing diversity. In this example therefore

$$1/D = 1/0.187 = 5.36$$

Source: James, F. C. and Shugart, H. H. (1970) A quantitative method of habitat description. *Audubon Field Notes*, December 1970, 727–36.

10 McIntosh's index of diversity

The McIntosh index of diversity is straightforward to calculate. In this example it is illustrated using data collected by Edwards and Brooker (1982) on the variety and abundance of macroinvertebrates in an upland section of the River Wye (UK). These data are shown in the table opposite.

The general form (U) of the McIntosh index is calculated from the following equation

$$U = \sqrt{(\Sigma n_i^2)}$$

where n_i is the proportional abundance of the ith species. The values of n_i^2 are shown in the data table.

Thus $$U = \sqrt{(\Sigma n_i^2)} = \sqrt{119812} = 346.14$$

This measure is strongly influenced by sample size. A dominance measure, which is independent of N (the total number of individuals), can be calculated using the formula

$$D = \frac{N - U}{N - \sqrt{N}} = \frac{1100 - 346.14}{1100 - \sqrt{1100}} = 0.7066$$

An additional evenness measure is obtained as follows

$$E = \frac{N - U}{N - N/\sqrt{S}} = \frac{1100 - 346.14}{1100 - 1100/\sqrt{38}} = 0.8180$$

Source: Edwards, R. W. and Brooker, M. P. (1982) *The Ecology of the Wye.* Junk, The Hague.

Species	*Number of individuals* n_i	n_i^2
Glossoma conformis	254	64516
Ephemeralla ignita	153	23409
Eiseniella tetraeda	90	8100
Simulium variegatum	69	4761
Simulium nitidifrons	68	4624
Simulium ornatum	58	3364
Baetis scambus	51	2601
Baetis rhodani	45	2025
Eusimulium aureum	40	1600
Limnius volckmari	39	1521
Simulium reptans	25	625
Dicranota sp. indet.	23	529
Thienemannimyia sp.	19	361
Enchytraedae indet.	18	324
Phagocata vitta	16	256
Hydropsyche siltalia	14	196
Rithrogena semicolorata	14	196
Rheotanytarsus sp. A.	11	121
Simulium reptans var. *galeratum*	11	121
Cricotopus sp. A.	11	121
Eukiefferiella verrallia	11	121
Lumbriculus variegatus	10	100
Eukiefferiella discoloripes	6	36
Ecdyonuruis dispar	6	36
Eukiefferiella clypeata	6	36
Cricotopus trifascia	6	36
Chloroperla tripunctata	5	25
Oulimnius tuberculatus	3	9
Baetis muticus	3	9
Elmis aenea	3	9
Esolus parallelepipedus	3	9
Nais alpina	3	9
Atherix ibis	1	1
Heptagenia sulphurea	1	1
Thienemanniella vittata	1	1
Hydra carina sp.	1	1
Cricotopus bicinctus	1	1
Rheocricotopus sp. indet.	1	1

Number of species $(S) = 38$
Total number of individuals $(N) = \Sigma n_i = 1100$
$\Sigma n_i^2 = 119\,812$

11 Berger–Parker diversity index

Wirjoatmodjo (1980) was interested in the feeding ecology of flounder (*Platichthys flesus*) in the estuary of the River Bann, Northern Ireland. He analysed the stomach contents of the fish at five sampling stations. The first of these was at the mouth of the river. Stations 2 and 3 were in the intertidal zone. Station 4 received sewage effluent while station 5 was subject to fresh water discharge from a weir and hot water discharge from a factory. The Berger–Parker index is employed in this example to determine whether there is any change in the dominance of food items in the flounder stomachs. The Berger–Parker index is calculated from the equation

$$d = N_{max}/N$$

where N = total number of individuals and N_{max} = number of individuals in the most abundant species. In order to ensure that the index increases with increasing diversity the reciprocal form of the measure is usually adopted.

	Number of individuals				
Food item	*Station 1*	*Station 2*	*Station 3*	*Station 4*	*Station 5*
Nereis	394	1642	90	126	32
Corophium	3487	5681	320	17	0
Gammarus	275	196	180	115	0
Tubifex	683	1348	46	436	5
Chironomid larvae	22	12	2	27	0
Other insect larvae	1	0	0	0	0
Arachnid	0	1	0	0	0
Carcinus	4	48	1	3	0
Cragnon	6	21	0	1	13
Neomysis	8	1	0	0	9
Sphaeroma	1	5	2	0	0
Flounder	1	7	1	1	0
Other fish	2	3	5	0	4
Number of species (S)	12	12	9	9	5
Number of individuals (N)	4884	8965	647	726	63
Most abundant species (N_{max})	3487	5681	320	436	32
Berger–Parker index					
$d = N_{max}/N$	0.714	0.634	0.495	0.601	0.508
$1/d$	1.40	1.58	2.02	1.67	1.96

The results shown in the table indicate that the greatest degree of dominance in food items occurs at the river mouth. Station 3 has the lowest dominance (and therefore highest evenness of food items). It is interesting to note that the greatest variety of food items occurs at Station 1.

Source: Wirjoatmodjo, S. (1980) Growth, food and movement of flounder (*Platichthys flesus L.*) in an estuary. Unpublished D. Phil. thesis, New University of Ulster.

12 Jack-knifing an index of diversity

As Chapter 2 pointed out, jack-knifing an index of diversity is a method of improving the estimate of virtually any statistic. In addition, it can be used to attach confidence limits to the estimate. Its main application in ecological diversity lies in situations where a number of samples have been taken. The basic technique involves recalculating overall diversity while missing out each sample in turn. Although the calculations are somewhat tedious (and a computer program is clearly desirable if it is used on a regular basis) the robustness of the method means that it should have increasingly wide application in investigations of ecological diversity. Virtually any diversity statistic can be employed. This example uses the reciprocal form of Simpson's index. (See Example 9 for details.) The data consist of the number of fish collected in five sections of the Upper Region of Black Creek, Mississippi (Ross *et al.*, 1987).

	Section					
Species	Σ	1	2	3	4	5
Esox americanus	14	13	0	0	1	0
Ericymba buccata	153	3	56	2	9	83
Notropis volucellus	261	38	77	4	31	111
N. venustus	1783	179	205	186	312	901
N. longirostris	100	4	0	6	1	89
N. texanus	1340	749	330	39	122	100
N. roseipinnis	4319	1827	918	173	945	456
Noturus leptacanthus	237	56	56	7	67	51
Labidesthes sicculus	163	145	4	0	7	7
Fundulus olivaceus	1075	585	123	130	190	47
Gambusia affinis	160	78	0	7	10	65
Aphredoderus sayanus	59	57	1	1	0	0
Ellassoma zonatum	54	43	5	0	4	2
Micropterus salmoides	38	20	4	0	3	11
Lepomis macrochirus	385	281	34	20	19	31
L. punctatus	26	26	0	0	0	0
L. megalotis	237	104	33	25	36	39
L. microlophus	36	23	0	2	4	7
L. cyanellus	36	23	1	7	5	0
Ammocrypta beani	280	60	72	105	30	13
Percina sciera	62	7	11	7	15	22
Ethostoma swaini	234	140	54	24	12	4
E. zonale	107	4	38	0	51	14
E. stigmaeum	201	39	52	40	46	24

The first step is to estimate the diversity of all stations together. Using Simpson's index (Example 9)

$$D_s = 4.96.$$

Then it is necessary to recalculate total diversity with each sample excluded. This will create five jack-knife estimates, VJ_i. Each of these jack-knife estimates is converted to a pseudovalue, VP_i, using the following equation

$$VP_i = (nV) - [(n-1)(VJ_i)]$$

where n = the number of samples.

The mean of the pseudovalues represents the best estimate of diversity (VP) and the difference between it and the initial estimate is a measure of what is called the 'sample influence function'.

In this example therefore the results are as follows.

Excluded section	VJ_i	VP_i
1	4.89	5.24
2	5.29	3.64
3	4.93	5.08
4	5.52	2.72
5	4.63	6.28

The mean of the VP_is is 4.59 and this is the best estimate (VP) of the diversity of fish in the river. Five samples is a rather small number from which to set confidence limits but where these are required they can be estimated in the usual way:

$$\text{Standard error of } VP = \text{Standard deviation of } VP_i\text{s}/\sqrt{(\text{no. of samples})}$$

Source: Ross, S. T., Baker, J. A. and Clark, K. E. (1987) Microhabitat partitioning of southeastern stream fishes: temporal and spatial predictability. In *Community and Evolutionary Ecology of North American Stream Fishes* (eds W. J. Matthews and D. C. Heins), University of Oklahoma Press, Norman and London, pp. 42–51.

13 Pielou's pooled quadrat method

Pielou's pooled quadrat method is a technique for estimating diversity when a random sample cannot be guaranteed. It involves repeatedly calculating the Brillouin index on randomly accumulated quadrats or samples. The diversity of the first sample is calculated, then the first two together, then the first three until all samples have been accounted for. The Brillouin cumulative diversity HB_k is plotted against the number of samples k. The point at which this curve flattens off is known as t and the flattened portion of the curve is used to estimate the population diversity HB_{pop}. To do this values of h_k from $k=t+1$ to $k=z$ (where $z=$ the total number of samples) are calculated from the formula

$$h_k = \frac{M_k HB_k - M_{k-1} HB_{k-1}}{M_k - M_{k-1}}$$

where $HB_k =$ the diversity of the kth (cumulative) sample calculated using the Brillouin index (see Example 8), and
$M_k =$ number of individuals (or other abundance measure) in the kth cumulative quadrat.

HB_{pop} is estimated by

$$HB_{pop} = \frac{1}{z-t+1} \sum h_k$$

The figure shows the cumulative diversity curve for the data collected in Breen Wood (see Example 6, page 142 for details). This curve is effectively flat at around 20 quadrats and the remaining 30 quadrats should therefore be used to estimate HB_{pop}. For simplicity however the process will be illustrated using cumulated quadrats 40 to 50.

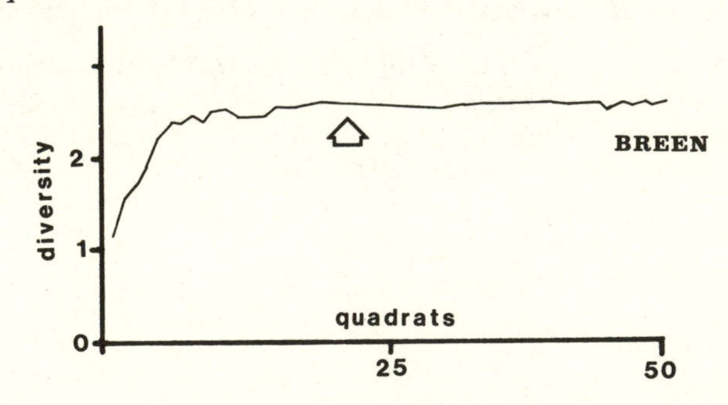

The table shows the Brillouin diversity (HB) calculated for cumulated quadrats 40 to 50 from Ness Wood. To facilitate calculations this table also incorporates values of M (total abundance), $M_k \times HB_k$, $M_k - M_{k-1}$ and $M_k \times HB_k - (M_{k-1} \times HB_{k-1})$.

k	HB	M	$M_k \times HB_k$	$M_k - M_{k-1}$	$M_k \times HB_k - (M_{k-1} \times HB_{k-1})$
40	2.57	718	1845.3		
41	2.56	731	1871.4	13	26.1
42	2.57	744	1912.1	13	40.7
43	2.55	764	1948.2	20	36.1
44	2.55	783	1996.7	19	48.5
45	2.54	794	2016.8	11	20.1
46	2.58	817	2107.9	23	91.1
47	2.57	827	2125.4	10	17.5
48	2.56	850	2176.0	23	50.6
49	2.57	865	2223.1	15	47.1
50	2.56	877	2245.1	12	22.1

Once these data have been assembled h_k is easily calculated. For instance for 41 quadrats

$$h_k = 26.1/13 = 2.01$$

This procedure is repeated until the 50 quadrat point is reached. The values of h_k are shown below.

k	h_k
41	2.01
42	3.13
43	1.81
44	2.55
45	1.83
46	3.96
47	1.75
48	2.20
49	3.14
50	1.84

The mean of these values is H_{pop} ($H_{pop} = 2.42$). Its standard deviation is simply the standard deviation of the values of h_k ($\sigma = 0.76$). This can be used to attach confidence limits in the usual way. For example 95% confidence limits are calculated from

$$t_{(9\ \mathrm{df})} \times \sigma/\sqrt{10} = 2.62 \times 0.24 = 0.63$$

14 β diversity

One purpose of β diversity measures is to ascertain the degree of turnover in species composition along a gradient or transect. This example deals with six measures used to calculate the β diversity on qualitative (that is presence and absence) data. For further information on these measures see page 91, Chapter 5. The data in the table are taken from an investigation of the vegetation of a nature reserve in Northern Ireland. They show the presence or absence of trees in six (10 m × 10 m) quadrats along a transect through a deciduous woodland.

	Transect						
Species	*1*	*2*	*3*	*4*	*5*	*6*	*Total occurrences*
Birch	×	×	×	—	—	—	3
Oak	×	×	×	×	×	×	6
Rowan	—	—	×	—	×	—	2
Beech	—	—	—	×	×	×	3
Hazel	—	—	—	—	×	×	2
Holly	—	—	—	×	—	×	2
Species	2	2	3	3	4	4	

The six measures discussed in Chapter 5 are calculated here.

1. The first of these is Whittaker's measure, β_w.

$$\beta_W = (S/\alpha) - 1$$

where S = the total number of species recorded in the system and α = the mean species richness.

$$\beta_W = 6/3 - 1 = 1$$

2. Cody's measure, β_c is the second index. It is calculated as

$$\beta_C = [g(H) + l(H)]/2$$

where $g(H)$ = the number of species gained along the transect and $l(H)$ = the number of species lost.

In this example two species (birch and oak) occur at the beginning of the transect. A further 4 are gained. Two species (birch and rowan) are lost at the end of the transect. The values in the equation are therefore

$$\beta_C = [4 + 2]/2 = 3$$

3, 4 and 5. Routledge proposed three measures of β diversity. The first of these, β_R, takes overall species richness and the degree of species overlap into consideration.

$$\beta_R = S^2/(2r+S)-1$$

where r = the number of species with overlapping distributions.

r can be calculated from a simple matrix which works out which pairs of species occur together in at least one quadrat.

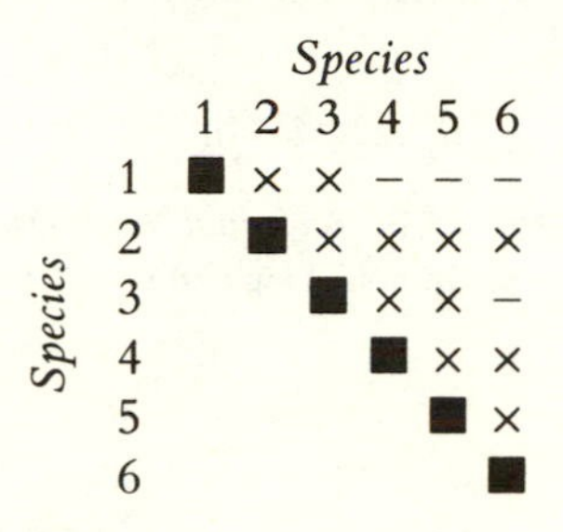

In this example there are 11 joint occurrences.

Thus $\beta_R = 6^2/(2\times 11+6)-1=(36/28)-1=0.2857$

Routledge's second measure, β_I, has its roots in information theory. It is calculated from the formula

$$\beta_I = \log(T) - [(1/T)\Sigma e_i \log(e_i)] - [(1/T)\Sigma \alpha_j \log(\alpha_j)]$$

where e_i is the number of samples along the transect in which species i is present, α_j is the species richness of sample j, and $T=\Sigma e_i=\Sigma \alpha_j$.

In order to be consistent with other diversity measures natural logs (ln) are adopted in this equation. The final column in the data table gives the number of samples in which each species occurs along the transect.

$\Sigma e_i \log(e_i)$ is therefore calculated as

$$(3\times \ln 3)+(6\times \ln 6)+ \cdots +(2\times \ln 2)=21.501$$

Similarly the final row in the data table gives the species richness of each quadrat.

$\Sigma \alpha_j \log(\alpha_j)$ is therefore calculated as

$$(2\times \ln 2)+(2\times \ln 2)+ \ldots +(4\times \ln 4)=20.454$$

$T=18$

Putting the equation together we get

$$\beta_I = \ln 18 - [(1/18)\times 21.501] - [(1/18)\times 20.454]$$
$$=0.5595$$

Routledge's final measure, β_E, is simply the exponential form of β_I

$$\beta_E = \exp(\beta_I) = \exp 0.5595 = 1.750$$

6. Wilson and Shmida's measure, β_T. This measure combines features of both Whittaker's measure and Cody's measure. It is calculated using the formula

$$\beta_T = [g(H) + 1(H)]/2\alpha$$

where α, $g(H)$ and $l(H)$ are defined as above.
In this example

$$\beta_T = [4 + 2]/6 = 1.00$$

Source: Magurran, A. E. (1976) *An Ecological Investigation of Boorin Nature Reserve.* Unpublished MS. Conservation Branch, Department of Environment, Northern Ireland.

15 Similarity measures

A further method of estimating β diversity employs similarity measures. This technique looks at the similarity of pairs of sites, either in terms of species presences and absences (qualitative data) or by taking species abundances into account (quantitative data). Although there are a vast range of these coefficients (see Chapter 5 for further details) this example restricts itself to four widely adopted measures. Two of these use presence and absence data while the other two require abundance data. The data in the table overleaf consist of the species (and abundances) of birds in managed and unmanaged areas along the River Wye (UK).

1. Jaccard measure (qualitative data)

This is calculated using the equation

$$C_J = j/(a+b-j)$$

where j = the number of species common to both sites
a = the number of species in site A, and
b = the number of species in site B.

Thus $C_J = 12/(26+12-12) = 0.46$

2. Sorenson measure (qualitative data)

This measure is similar to the Jaccard index and uses identical variables.

$$C_S = 2j/(a+b)$$

Thus $$C_S = 24/(26+12) = 0.63$$

3. Sorenson measure (quantitative data)

A version of the Sorenson measure uses quantitative data. The formula is

$$C_N = 2_{jN}/(aN+bN)$$

where aN = the number of individuals in site A, bN = the number of individuals in site B, and jN = the sum of the lower of the two abundances of species which occur in the two sites. In this example jN is therefore $(2.9+10+5.7+\ldots+2.9) = 58.4$. It is identical to the sum of the abundances in the managed area because abundances of the bird species are always lowest in this habitat.

Thus $$C_N = 2 \times 58.4/(204.5+58.4) = 0.44$$

Species	Territories per 10 km	
	Unmanaged	Managed
Great-crested grebe	1.4	0
Mallard	4.3	0
Mute swan	2.9	0
Moorhen	8.6	2.9
Coot	4.2	0
Common sandpiper	15.7	0
Kingfisher	2.0	0
Sandmartin	50	10
Dipper	1	0
Sedge warbler	11.4	0
Pied wagtail	11.4	5.7
Grey wagtail	4.3	2.5
Yellow wagtail	13.0	5.7
Reed bunting	14.3	8.6
Heron	8.6	5.7
Curlew	7.1	2.9
Lapwing	10.0	0
Redshank	1.4	0
Nuthatch	2.9	2.9
Tree-creeper	5.7	0
Whinchat	1.4	0
Blackcap	11.4	5.7
Garden warbler	2.9	0
Whitethroat	4.3	2.9
Lesser whitethroat	1.4	0
Spotted fly-catcher	2.9	2.9
Number of species (S)	26	12
Total number of individuals (N)	204.5	58.4

Note: Abundance in this example is strictly speaking the number of territories. The phrase 'number of individuals' is however retained as the general term for abundance in order to maintain consistent terminology throughout the book.

4. Morista–Horn measure (quantitative data)

This index is calculated from the equation

$$C_{\mathrm{MH}} = \frac{2\Sigma(an_i \times bn_i)}{(da + db)aN \times bN}$$

where aN = the number of individuals in site A,
bN = the number of individuals in site B,
an_i = the number of individuals in the ith species in site A,
bn_i = the number of individuals in the ith species in site B

$$da = \frac{\Sigma an_i^2}{aN^2} \text{ and } db = \frac{\Sigma bn_i^2}{bN^2}$$

In this example therefore

$$\Sigma(an_i \times bn_i) = (1.4 \times 0 + 4.3 \times 0 + 2.9 \times 0 + 8.6 \times 2.9 + \cdots + 2.9 \times 2.9)$$

$$= 961.63$$

$$da = \frac{\Sigma an_i^2}{aN^2} = \frac{3960.27}{41820.25} = 0.0947$$

and $$db \frac{\Sigma bn_i^2}{bN^2} = \frac{352.22}{3410.56} = 0.1033$$

Thus $$C_{\mathrm{MH}} = \frac{2\Sigma(an_i \times bn_i)}{(da + db)aN \times bN} = \frac{2 \times 961.63}{(0.0947 + 0.1033)\,(204.5 \times 58.4)}$$

$$= \frac{1923.26}{2364.67} = 0.8133$$

Source: Edwards, R. W. and Brooker, M. P. (1982) *The Ecology of the Wye.* Junk, The Hague.

Appendix
Statistics tables

A1 Distribution of χ^2

Degrees of freedom	Probability													
	.99	.98	.95	.90	.80	.70	.50	.30	.20	.10	.05	.02	.01	.001
1	$.0^3157$	$.0^3628$	.00393	.0158	.0642	.148	.455	1.074	1.642	2.706	3.841	5.412	6.635	10.827
2	.0201	.0404	.103	.211	.446	.713	1.386	2.408	3.219	4.605	5.991	7.824	9.210	13.815
3	.115	.185	.352	.584	1.005	1.424	2.366	3.665	4.642	6.251	7.815	9.837	11.341	16.268
4	.297	.429	.711	1.064	1.649	2.195	3.357	4.878	5.989	7.779	9.488	11.668	13.277	18.465
5	.554	.752	1.145	1.610	2.343	3.000	4.351	6.064	7.289	9.236	11.070	13.388	15.086	20.517
6	.872	1.134	1.635	2.204	3.070	3.828	5.348	7.231	8.558	10.645	12.592	15.033	16.812	22.457
7	1.239	1.564	2.167	2.833	3.822	4.671	6.346	8.383	9.803	12.017	14.067	16.622	18.475	24.322
8	1.646	2.032	2.733	3.490	4.594	5.527	7.344	9.524	11.030	13.362	15.507	18.168	20.090	26.125
9	2.088	2.532	3.325	4.168	5.380	6.393	8.343	10.656	12.242	14.684	16.919	19.679	21.666	27.877
10	2.558	3.059	3.940	4.865	6.179	7.267	9.342	11.781	13.442	15.987	18.307	21.161	23.209	29.588
11	3.053	3.609	4.575	5.578	6.989	8.148	10.341	12.899	14.631	17.275	19.675	22.618	24.725	31.264
12	3.571	4.178	5.226	6.304	7.807	9.034	11.340	14.011	15.812	18.549	21.026	24.054	26.217	32.909
13	4.107	4.765	5.892	7.042	8.634	9.926	12.340	15.119	16.985	19.812	22.362	25.472	27.688	34.528
14	4.660	5.368	6.571	7.790	9.467	10.821	13.339	16.222	18.151	21.064	23.685	26.873	29.141	36.123
15	5.229	5.985	7.261	8.547	10.307	11.721	14.339	17.322	19.311	22.307	24.996	28.259	30.578	37.697
16	5.812	6.614	7.962	9.312	11.152	12.624	15.338	18.418	20.465	23.542	26.296	29.633	32.000	39.252
17	6.408	7.255	8.672	10.085	12.002	13.531	16.338	19.511	21.615	24.769	27.587	30.995	33.409	40.790
18	7.015	7.906	9.390	10.865	12.857	14.440	17.338	20.601	22.760	25.989	28.869	32.346	34.805	42.312
19	7.633	8.567	10.117	11.651	13.716	15.352	18.338	21.689	23.900	27.204	30.144	33.687	36.191	43.820
20	8.260	9.237	10.851	12.443	14.578	16.266	19.337	22.775	25.038	28.412	31.410	35.020	37.566	45.315
21	8.897	9.915	11.591	13.240	15.445	17.182	20.337	23.858	26.171	29.615	32.671	36.343	38.932	46.797
22	9.542	10.600	12.338	14.041	16.314	18.101	21.337	24.939	27.301	30.813	33.924	37.659	40.289	48.268
23	10.196	11.293	13.091	14.848	17.187	19.021	22.337	26.018	28.429	32.007	35.172	38.968	41.638	49.728
24	10.856	11.992	13.848	15.659	18.062	19.943	23.337	27.096	29.553	33.196	36.415	40.270	42.980	51.179
25	11.524	12.697	14.611	16.473	18.940	20.867	24.337	28.172	30.675	34.382	37.652	41.566	44.314	52.620
26	12.198	13.409	15.379	17.292	19.820	21.792	25.336	29.246	31.795	35.563	38.885	42.856	45.642	54.052
27	12.879	14.125	16.151	18.114	20.703	22.719	26.336	30.319	32.912	36.741	40.113	44.140	46.963	55.476
28	13.565	14.847	16.928	18.939	21.588	23.647	27.336	31.391	34.027	37.916	41.337	45.419	48.278	56.893
29	14.256	15.574	17.708	19.768	22.475	24.577	28.336	32.461	35.139	39.087	42.557	46.693	49.588	58.302
30	14.953	16.306	18.493	20.599	23.364	25.508	29.336	33.530	36.250	40.256	43.773	47.962	50.892	59.703

A2 Distribution of t

Degrees of freedom	Probability					
	.20	.10	.05	.02	.01	.001
1	3.078	6.314	12.706	31.821	63.657	636.619
2	1.886	2.920	4.303	6.965	9.925	31.598
3	1.638	2.353	3.182	4.541	5.841	12.941
4	1.533	2.132	2.776	3.747	4.604	8.610
5	1.476	2.015	2.571	3.365	4.032	6.859
6	1.440	1.943	2.447	3.143	3.707	5.959
7	1.415	1.895	2.365	2.998	3.499	5.405
8	1.397	1.860	2.306	2.896	3.355	5.041
9	1.383	1.833	2.262	2.821	3.250	4.781
10	1.372	1.812	2.228	2.764	3.169	4.587
11	1.363	1.796	2.201	2.718	3.106	4.437
12	1.356	1.782	2.179	2.681	3.055	4.318
13	1.350	1.771	2.160	2.650	3.012	4.221
14	1.345	1.761	2.145	2.624	2.977	4.140
15	1.341	1.753	2.131	2.602	2.947	4.073
16	1.337	1.746	2.120	2.583	2.921	4.015
17	1.333	1.740	2.110	2.567	2.898	3.965
18	1.330	1.734	2.101	2.552	2.878	3.922
19	1.328	1.729	2.093	2.539	2.861	3.883
20	1.325	1.725	2.086	2.528	2.845	3.850
21	1.323	1.721	2.080	2.518	2.831	3.819
22	1.321	1.717	2.074	2.508	2.819	3.792
23	1.319	1.714	2.069	2.500	2.807	3.767
24	1.318	1.711	2.064	2.492	2.797	3.745
25	1.316	1.708	2.060	2.485	2.787	3.725
26	1.315	1.706	2.056	2.479	2.779	3.707
27	1.314	1.703	2.052	2.473	2.771	3.690
28	1.313	1.701	2.048	2.467	2.763	3.674
29	1.311	1.699	2.045	2.462	2.756	3.659
30	1.310	1.697	2.042	2.457	2.750	3.646
40	1.303	1.684	2.021	2.423	2.704	3.551
60	1.296	1.671	2.000	2.390	2.660	3.460
120	1.289	1.658	1.980	2.358	2.617	3.373
∞	1.282	1.645	1.960	2.326	2.576	3.291

A3 Cohen's table for truncated log normal

γ	.000	.001	.002	.003	.004	.005	.006	.007	.008	.009	γ
0.005	.00000	.00000	.00000	.00001	.00001	.00001	.00001	.00001	.00002	.00002	0.05
0.06	.00002	.00003	.00003	.00003	.00004	.00004	.00005	.00006	.00007	.00007	0.06
0.07	.00008	.00009	.00010	.00011	.00013	.00014	.00016	.00017	.00019	.00020	0.07
0.08	.00022	.00024	.00026	.00028	.00031	.00033	.00036	.00039	.00042	.00045	0.08
0.09	.00048	.00051	.00055	.00059	.00063	.00067	.00071	.00075	.00080	.00085	0.09
0.10	.00090	.00095	.00101	.00106	.00112	.00118	.00125	.00131	.00138	.00145	0.10
0.11	.00153	.00160	.00168	.00176	.00184	.00193	.00202	.00211	.00220	.00230	0.11
0.12	.00240	.00250	.00261	.00272	.00283	.00294	.00305	.00317	.00330	.00342	0.12
0.13	.00355	.00369	.00382	.00396	.00410	.00425	.00440	.00455	.00470	.00486	0.13
0.14	.00503	.00519	.00536	.00553	.00571	.00589	.00608	.00627	.00646	.00665	0.14
0.15	.00685	.00705	.00726	.00747	.00769	.00791	.00813	.00835	.00858	.00882	0.15
0.16	.00906	.00930	.00955	.00980	.01006	.01032	.01058	.01085	.01112	.01140	0.16
0.17	.01168	.01197	.01226	.01256	.01286	.01316	.01347	.01378	.01410	.01443	0.17
0.18	.01476	.01509	.01543	.01577	.01611	.01646	.01682	.01718	.01755	.01792	0.18
0.19	.01830	.01868	.01907	.01946	.01986	.02026	.02067	.02108	.02150	.02193	0.19
0.20	.02236	.02279	.02323	.02368	.02413	.02458	.02504	.02551	.02599	.02647	0.20
0.21	.02695	.02744	.02794	.02844	.02895	.02946	.02998	.03050	.03103	.03157	0.21
0.22	.03211	.03266	.03322	.03378	.03435	.03492	.03550	.03609	.03668	.03728	0.22
0.23	.03788	.03849	.03911	.03973	.04036	.04100	.04165	.04230	.04296	.04362	0.23
0.24	.04429	.04497	.04565	.04634	.04704	.04774	.04845	.04917	.04989	.05062	0.24
0.25	.05136	.05211	.05286	.05362	.05439	.05516	.05594	.05673	.05753	.05834	0.25
0.26	.05915	.05997	.06080	.06163	.06247	.06332	.06418	.06504	.06591	.06679	0.26
0.27	.06768	.06858	.06948	.07039	.07131	.07224	.07317	.07412	.07507	.07603	0.27
0.28	.07700	.07797	.07896	.07995	.08095	.08196	.08298	.08401	.08504	.08609	0.28
0.29	.08714	.08820	.08927	.09035	.09144	.09254	.09364	.09476	.09588	.09701	0.29
0.30	.09815	.09930	.10046	.10163	.10281	.10400	.10520	.10641	.10762	.10885	0.30
0.31	.1101	.1113	.1126	.1138	.1151	.1164	.1177	.1190	.1203	.1216	0.31
0.32	.1230	.1243	.1257	.1270	.1284	.1298	.1312	.1326	.1340	.1355	0.32
0.33	.1369	.1383	.1398	.1413	.1428	.1443	.1458	.1473	.1488	.1503	0.33
0.34	.1519	.1534	.1550	.1566	.1582	.1598	.1614	.1630	.1647	.1663	0.34
0.35	.1680	.1697	.1714	.1731	.1748	.1765	.1782	.1800	.1817	.1835	0.35
0.36	.1853	.1871	.1899	.1907	.1926	.1944	.1963	.1982	.2001	.2020	0.36
0.37	.2039	.2058	.2077	.2097	.2117	.2136	.2156	.2176	.2197	.2217	0.37
0.38	.2238	.2258	.2279	.2300	.2321	.2342	.2364	.2385	.2407	.2429	0.38
0.39	.2451	.2473	.2495	.2517	.2540	.2562	.2585	.2608	.2631	.2655	0.39
0.40	.2678	.2702	.2726	.2750	.2774	.2798	.2822	.2847	.2871	.2896	0.40
0.41	.2921	.2947	.2972	.2998	.3023	.3049	.3075	.3102	.3128	.3155	0.41
0.42	.3181	.3208	.3235	.3263	.3290	.3318	.3346	.3374	.3402	.3430	0.42
0.43	.3459	.3487	.3516	.3545	.3575	.3604	.3634	.3664	.3694	.3724	0.43

0.44	.3755	.3785	.3816	.3847	.3878	.3910	.3941	.3973	.4005	.4038	0.44
0.45	.4070	.4103	.4136	.4169	.4202	.4236	.4269	.4303	.4338	.4372	0.45
0.46	.4407	.4442	.4477	.4512	.4547	.4583	.4619	.4655	.4692	.4728	0.46
0.47	.4765	.4802	.4840	.4877	.4915	.4953	.4992	.5030	.5069	.5108	0.47
0.48	.5148	.5187	.5227	.5267	.5307	.5348	.5389	.5430	.5471	.5513	0.48
0.49	.5555	.5597	.5639	.5682	.5725	.5768	.5812	.5856	.5900	.5944	0.49
0.50	.5989	.6034	.6079	.6124	.6170	.6216	.6263	.6309	.6356	.6404	0.50
0.51	.6451	.6499	.6547	.6596	.6645	.6694	.6743	.6793	.6843	.6893	0.51
0.52	.6944	.6995	.7046	.7098	.7150	.7202	.7255	.7308	.7361	.7415	0.52
0.53	.7469	.7524	.7578	.7633	.7689	.7745	.7801	.7857	.7914	.7972	0.53
0.54	.8029	.8087	.8146	.8204	.8263	.8323	.8383	.8443	.8504	.8565	0.54
0.55	.8627	.8689	.8751	.8813	.8876	.8940	.9004	.9068	.9133	.9198	0.55
0.56	.9264	.9330	.9396	.9463	.9530	.9598	.9666	.9735	.9804	.9874	0.56
0.57	.9944	1.001	1.009	1.016	1.023	1.030	1.037	1.045	1.052	1.060	0.57
0.58	1.067	1.075	1.082	1.090	1.097	1.105	1.113	1.121	1.129	1.137	0.58
0.59	1.145	1.153	1.161	1.169	1.177	1.185	1.194	1.202	1.211	1.219	0.59
0.60	1.228	1.236	1.245	1.254	1.262	1.271	1.280	1.289	1.298	1.307	0.60
0.61	1.316	1.326	1.335	1.344	1.353	1.363	1.373	1.382	1.392	1.402	0.61
0.62	1.411	1.421	1.431	1.441	1.451	1.461	1.472	1.482	1.492	1.503	0.62
0.63	1.513	1.524	1.534	1.545	1.556	1.567	1.578	1.589	1.600	1.611	0.63
0.64	1.622	1.634	1.645	1.657	1.668	1.680	1.692	1.704	1.716	1.728	0.64
0.65	1.740	1.752	1.764	1.777	1.789	1.802	1.814	1.827	1.840	1.853	0.65
0.66	1.866	1.879	1.892	1.905	1.919	1.932	1.946	1.960	1.974	1.988	0.66
0.67	2.002	2.016	2.030	2.044	2.059	2.073	2.088	2.103	2.118	2.133	0.67
0.68	2.148	2.163	2.179	2.194	2.210	2.225	2.241	2.257	2.273	2.290	0.68
0.69	2.306	2.322	2.339	2.356	2.373	2.390	2.407	2.424	2.441	2.459	0.69
0.70	2.477	2.495	2.512	2.531	2.549	2.567	2.586	2.605	2.623	2.643	0.70
0.71	2.662	2.681	2.701	2.720	2.740	2.760	2.780	2.800	2.821	2.842	0.71
0.72	2.863	2.884	2.905	2.926	2.948	2.969	2.991	3.013	3.036	3.058	0.72
0.73	3.081	3.104	3.127	3.150	3.173	3.197	3.221	3.245	3.270	3.294	0.73
0.74	3.319	3.344	3.369	3.394	3.420	3.446	3.472	3.498	3.525	3.552	0.74
0.75	3.579	3.606	3.634	3.662	3.690	3.718	3.747	3.776	3.805	3.834	0.75
0.76	3.864	3.894	3.924	3.955	3.986	4.017	4.048	4.080	4.112	4.144	0.76
0.77	4.177	4.210	4.243	4.277	4.311	4.345	4.380	4.415	4.450	4.486	0.77
0.78	4.52	4.56	4.60	4.63	4.67	4.71	4.75	4.79	4.82	4.86	0.78
0.79	4.90	4.94	4.99	5.03	5.07	5.11	5.15	5.20	5.24	5.28	0.79
0.80	5.33	5.37	5.42	5.46	5.51	5.56	5.61	5.65	5.70	5.75	0.80
0.81	5.80	5.85	5.90	5.95	6.01	6.06	6.11	6.17	6.22	6.28	0.81
0.82	6.33	6.39	6.45	6.50	6.56	6.62	6.68	6.74	6.81	6.87	0.82
0.83	6.93	7.00	7.06	7.13	7.19	7.26	7.33	7.40	7.47	7.54	0.83
0.84	7.61	7.68	7.76	7.83	7.91	7.98	8.06	8.14	8.22	8.30	0.84
0.85	8.39	8.47	8.55	8.64	8.73	8.82	8.91	9.00	9.09	9.18	0.85

Index

Numbers in *italics* indicate figures or tables